Selected Titles in This Series

University
LECTURE
Series

Volume 11

Some Points of Analysis and Their History

Lars Gårding

American Mathematical Society
Providence, Rhode Island

Higher Education Press
Beijing, China

DTF

1991 *Mathematics Subject Classification.* Primary 01A72;
Secondary 30–03, 35–03, 42–03, 46–03.

ABSTRACT. The purpose of these lectures is to give historical background and leisurely accounts of some important results in analysis in this century. Most of them belong to the classical analysis and the theory of partial differential operators and are associated with Swedish mathematicians, but there is also the Tarski-Seidenberg theorem and Wiener's classical results in harmonic analysis.

The book is of interest to all specialists and students in analysis and partial differential equations.

Library of Congress Cataloging-in-Publication Data
Gårding, Lars, 1919–
  Some points of analysis and their history / Lars Gårding.
    p. cm. — (University lecture series ; v. 11)
  Includes bibliographical references (p. –).
  ISBN 0-8218-0757-9 (acid-free paper)
  1. Mathematical analysis. I. Title. II. Series: University lecture series (Providence, R. I.) ;
11.
QA300.5.G37   1997
515–dc21

97-24642
CIP

NWST
1ALD9690

Contents

Author's Preface

The purpose of these lectures is to give historical background and leisurely accounts of some important results in analysis in this century. Most of the results in classical analysis and the theory of partial differential operators are associated with Swedish mathematicians, but we also include the Tarski-Seidenberg theorem and Wiener's classical results in harmonic analysis, which have demonstrated over time that simple things may lie behind problems that were once very famous and that engendered much work.

It goes without saying that the circle of problems treated here represent just a tiny fraction of the thousands of important results in analysis. Personal affinity rather than systematic selection has determined my sample.

The inspiration for the lectures was an invitation to join the centenary of Wuhan University in 1993. For various reasons I was not able to attend at that time, but I gave some of the lectures when I visited Nankai, Wuhan, Fudan, Jilin and Beijing universities a year later. I want to express my gratitude for the courtesy extended to me by all these universities.

I also thank Professor Li Da-Tsien for arranging the printing of my lectures and their translation into Chinese. The present expanded version, including lectures on Picard's great theorem, Nevanlinna theory, and a personal essay on the impact of distributions in analysis, has been accepted by the American Mathematical Society.[1] Finally, I thank Jana Madjarova for careful proofreading, Natalya Pluzhnikov for expert editing, and Sven Spanne for helping me with a tricky font.

Lars Gårding

Lund, 1997

---

[1] The original title, "Some problems of analysis and their history", has now been changed to "Some points of analysis and their history".

CHAPTER 1

Picard's Great Theorem

Introduction

Charles Émile Picard (1856–1941) is famous for Picard's theorem. In its general form it says that a meromorphic function assumes all but two values in any neighborhood of an isolated essentially singular point. Note that a meromorphic function may have poles of finite order and that these are not essential singularities.

In all proofs the assumption that f avoids three different values in a neighborhood of an isolated essentially singular point leads to a contradiction. Picard's own proof was an unexpected fruit of the theory of elliptic functions. His tool was the so-called modular function, the result of half a century of intense study.

Picard's theorem was a radical improvement on Weierstrass's result that an analytic function comes arbitrarily close to any given value in any neighborhood of an isolated essentially singular point, but the theorem seemed mysterious for several reasons. The use of modular functions was out of proportion with the simple formulation of the theorem and gave no hint why precisely two exceptional values were the maximum. In the many later proofs of Picard's theorem the modular function was first eliminated by the use of various inequalities by Borel (1897) and Schottky (1904). Landau (1916, 1929) gave a terse account of Schottky's inequality in his classic *Neuere Ergebnisse der Funktionentheorie*. Rolf Nevanlinna's book (1929) was motivated by Picard's theorem and contains it as a special case of a general theory of the exceptional values of functions which are meromorphic outside a compact set, the point infinity excepted. Nevanlinna's theory will be sketched in Chapter 4. Finally, in (1935b) Lars Ahlfors gave a topological explanation why there are at most two exceptional values.

Tha aim of this paper is to present or at least sketch some of the proofs in this turn of events, starting with Picard's own proof.

Picard's proof

The task is to prove that a meromorphic function cannot avoid three values in any neighborhhod of an isolated essentially singular point. Here the exceptional values may be taken to be $0, 1, \infty$ so that it suffices to consider an analytic function at an isolated essentially singular point which does not take the values 0 and 1. In fact, if the values not taken are a, b, c, the function

$$g(z) = \frac{f(z) - a}{f(z) - b} : \frac{c - a}{c - b}$$

avoids the values $0, 1, \infty$, even if one of a, b, c already is infinity, and if f has an isolated essential singularity at z_0, so does g and conversely.

Picard proved two versions of his theorem. The first one (1879) says that an entire function whose range avoids two separate complex numbers is a constant. The main theorem was proved one year later (1880). In his papers he could just refer to known properties of modular functions. For completeness we shall now describe the one he used.

The modular function

The elliptic integral

$$u = \int_0^y \frac{1}{\sqrt{(1 - t^2)(1 - \kappa^2 t^2)}} dt$$

where the *module* κ^2 is not $0, 1, \infty$, defines Jacobi's elliptic function $y = s(\kappa, u)$, the *sinus amplitudinis*. For $0 < \kappa^2 < 1$ it has two canonically defined periods σ_1, σ_2 obtained by integration along certain closed cycles on the Riemann surface of the curve $y^2 = (1 - x^2)(1 - \kappa^2 x^2)$. All other periods are then linear combinations of these two periods, which can always be chosen so that the quotient $\omega = \sigma_1/\sigma_2$ has a positive imaginary part.

When the module $z = \kappa^2$ avoids the values $0, 1, \infty$, the quotient $\omega = \omega(z)$ is an analytic many-valued function of $z = \kappa^2$ with values in the upper half-plane. Under closed loops, $\omega(z)$ is subject to certain Möbius transformations which form a discrete group Γ. More precisely, the images under ω of the lower and upper half-planes form a tesselation of the upper half-plane by non-Euclidean triangles with all three corners on the real axis. The inverse of ω is a function from the upper half-plane to itself which is automorphic in the sense that it is invariant under Γ.

Picard's two papers

In his first paper Picard used only the fact that the quotient $\omega = \omega(z)$ is analytic and many-valued with values in the upper half-plane when z is not equal to $0, 1, \infty$. After a Möbius map which makes an entire function $f(z)$ avoid the points $0, 1$, the proof of Picard's first theorem is now obvious: $\omega(f(z))$ can be continued analytically everywhere in the complex plane, hence it is a single-valued entire function with range in the upper half-plane and must be constant so that f is constant.

Very soon afterwards Picard could prove also his second theorem by a variation of the same trick. Actually, Picard's proof is elementary modulo the existence of an automorphic function defined in the upper half-plane. It is difficult to read only because the author uses the theory of Möbius maps in a complicated way. This was before the present canonical theory of linear algebra, and it is hoped that the rendering of the proof below is more readable. For completeness the text includes a sketch of the classical construction of automorphic functions.

Möbius maps

A Möbius map is an invertible fractional linear map

$$z \rightarrow \frac{az + b}{cz + d}, \quad ad - bc = 1,$$

with complex coefficients. When A is the matrix $(a, b)/(c, d)$, it is convenient to write the right side above as

$$A[z] = \frac{az + b}{cz + d}.$$

It is immediate to verify that A is invertible and that

$$AB[z] = (AB)[z].$$

Reflection $z \to w$ in a circle with center α and radius r is given by the formula

$$\overline{(z - \alpha)}(w - \alpha) = r^2,$$

and it is known by elementary geometry that reflections map circles to circles. Our map may be written as $w = A[\bar{z}]$ with a certain invertible matrix A. Hence every reflection in circles (including straight lines) is an improper Möbius map, that is, a Möbius map preceeded or followed by a conjugation. The product of two such maps is a Möbius map. In fact, all proper and improper Möbius maps form a group, the full Möbius group M, generated by reflections. All its elements map circles into circles.

Tesselations of the upper half-plane. Automorphic functions

A triangle bounded by circular arcs which touch each other at the corners so that all corner angles vanish needs a simple name. Let us call it a vanishing triangle. Automorphic functions are closely connected with tesselations of the upper half-plane H by vanishing triangles with all corners on the real axis. Such a tesselation may start with a triangle K in H bounded by the lines $x = 0$ and $x = 1$ and the half-circle $|z - 1/2| = 1/2$. Reflections in the sides will then give three adjoining vanishing triangles whose sides like those of K meet the real axis under right angles and belong to H. Repeated reflections will then produce a tesselation of the upper half-plane.[1] At the same time they generate a subgroup G of the full Möbius group which maps the tesselation to itself. This group is discrete in the sense that if $A, B \in G$ and $z \in H$, then $A = B$ when Az and Bz are sufficiently close.

We can now construct automorphic functions simply by using Riemann's mapping theorem to map K conformally to the upper half-plane H by a function φ which maps the corners $0, 1, \infty$ to themselves. If a reflection R maps K to any of its neighbors, then, by Schwarz's reflection principle, $\varphi(z) = \overline{\varphi(R^{-1}z)}$ extends φ to RK across their common boundary in such a way that RK is mapped to the lower half-plane. Continuing this process we have a function $I(z)$ defined in the upper half-plane and invariant under G. The corners of the tesselation are all mapped to one of the points $0, 1, \infty$.

The inverse function $J(z)$ is many-valued but has the crucial property of being singular only at the points $0, 1, \infty$. When z runs through a closed path γ from $z_0 \in H$ back to z_0 which avoids these points, then $J(z)$ assumes a new value $A[J(z_0)]$ for some $A \in G$. Since γ crosses the real line an even number of times, we are sure that A is a Möbius map.

---

[1] The reader is advised to draw a figure himself or to look up a corresponding figure in some standard treatise.

Normal forms of Möbius maps of H

In our version of Picard's proof of his theorem, we shall need to know the normal forms of Möbius maps $z \to A[z]$ of H to itself. Then the real axis is mapped to itself so that A may be assumed to be real. Further, since

$$\text{Im}\,\frac{az + b}{cz + d} = (ad - bc)\frac{\text{Im}\,z}{|cz + d|^2},$$

the determinant $ad - bc$ is positive. We normalize it to 1. Then the eigenvalues of A have the form $\lambda, 1/\lambda$ and, since their sum is real, they are either both real or both of absolute value 1.

The possible normal forms of A under similarity maps $A \to SAS^{-1}$ are as follows:

1. Two complex eigenvalues, S maps H to the unit circle and $SAS^{-1} = D$ is diagonal with non-real elements $e^{i\theta}, e^{-i\theta}$.

2. Two real eigenvalues $\lambda > 1$ and $1/\lambda$, S maps H to itself, SAS^{-1} is diagonal with elements $\lambda, 1/\lambda$.

3. Two eigenvalues equal to 1, A is not diagonalizable, but there is an S which maps H to itself such that SAS^{-1} is the matrix $(1, 1)/(0, 1)$.

4. A is the unit matrix.

Proof of Picard's theorem

Using the elementary statements of the previous section we can now prove the

THEOREM. *A function $f(z)$ which is analytic and single-valued when restricted to a neighborhood N of ∞ can avoid at most one value.*

In the proof we may assume that $f(z)$ is never 0 or 1 and use the inverse $J(z)$ of the automorphic function $I(z)$ defined above. The following lemma is taken as a matter of course by Picard.

LEMMA. *With f as above, the function $g(z) = J(f(z))$ is analytic in a connected neighborhood N of ∞ with values in H and, under a turn T in N in positive direction around the origin,*

$$(1) \qquad\qquad Tg(z) = A[g(z)]$$

for some Möbius map $A \in G$.

PROOF. Since $f(z)$ is never $0, 1, \infty$, $g(z)$ can be continued analytically and indefinitely in N. Also, if T refers to a path $\gamma \subset N$ in positive direction from a point z_0 and back, it is clear that (1) holds with $z = z_0$ and some $A \in G$. A slight modification of γ will change A to some $A' \in G$ close to A, but since G is discrete, $A'z_0$ cannot come arbitrarily close to Az_0 unless $A' = A$. Hence A does not depend on the choice of γ. Similarly, it cannot depend on the choice of z_0. The last statement follows from Weierstrass's theorem.

In the rest of the proof we shall see that (1) leads to situations where the range of $f(z)$ for large z cannot be dense in the upper half-plane contradicting Weierstrass's theorem.

1. Suppose that A has complex eigenvalues $e^{i\theta}, e^{-i\theta}$ and let S be a diagonalizing matrix mapping H to the unit disk. We may assume that $0 < \theta < \pi$. Then

$$S[Tg(z)] = e^{2i\theta}S[g(z)]$$

so that

$$S[g(z)] = z^{\theta/\pi}h(z)$$

where $h(z)$ is single-valued and $|S[g(z)]| \leq 1$. This is possible only if $h(z) = O(1/z)$ and then the left side tends to zero as $z \to \infty$. But then $g(z) = J(f(z))$ has a limit in the upper half-plane as $z \to \infty$ so that f cannot be singular at infinity.

2. Suppose that A has real eigenvalues $\lambda > 1$ and $1/\lambda$ so that $SH \subset H$. Now $z^{\log \lambda/\pi i}$ changes by a factor of λ^2 under T and hence

$$S[T^n g(z)] = z^{n \log \lambda/\pi i}h(z)$$

where $h(z)$ is single-valued and the left side belongs to H. Here, since $\log \lambda > 0$, we can put $z = e^{m/\log \lambda}$ where $m > 0$ is a large integer and then the range of

$$e^{nr \log \lambda/\pi i} = e^{nm/\pi i}$$

is dense in the unit circle when n varies. Hence $g(z)$ is not in H and this is a contradiction.

3. We may suppose that $SA[w] = w + 1$ and that $SH \subset H$. Then

(2) $$S[Tg(z)] = \frac{\log z}{2\pi i} + h(z)$$

for all n where $h(z)$ is analytic and single-valued for large arguments. Hence

$$e^{iS[Tg(z)]} = z^{1/2\pi}e^{ih(z)}$$

where the left side is bounded and its range dense in the unit disk for all regions $|z| > \text{const}$. But then $e^{ih(z)}$ tends to zero at least as $1/z$ and this is a contradiction.

4. Suppose that $g(z)$ is single-valued. Then this function cannot have values in H unless it is regular at infinity, and this means that $g(z)$ tends to a limit as $z \to \infty$. But the range of $g(z) = J(f(z))$ is dense in the range of J in every neighborhood of ∞, whence a contradiction.

The proofs by Borel and Schottky

Picard's proof of Picard's theorem explores the absurd consequences of the assumption that there exists an entire function which avoids two separate values or the absurd consequences of the existence of a function which avoids three values in the neighborhood of an isolated essential singularity and is meromorphic outside. The ensuing proofs of Borel (1897) and Schottky (1904) avoid the theory of elliptic functions. Borel's proof, which is simple only in principle, only concerns Picard's first theorem about entire functions.

In the first edition of his classic *Leçons sur les fonctions entières* (1900) Émile Borel devoted a chapter to Picard's theorem and almost proved its analytic version by a very simple argument using the concept of growth of entire functions. A simple paraphrase of Borel's argument runs as follows.

As has been remarked before, it suffices to consider an entire function $f(z)$ which is never 0 or 1. Then $f(z) = e^{g(z)}$ for some entire function $g(z)$ never equal to an integral multiple of $2\pi i$ and hence also

$$f(z) = e^{-2\pi i g(z)}$$

where g does not take integral values, in particular not 0 or 1. Hence, if

$$M(f,r) = \max_{|z|=r} |f(z)|, \quad A(f,r) = \max_{|z|=r} \operatorname{Im} f(z),$$

we must have

$$M(f,r) \le e^{2\pi A(g,r)}.$$

Now the value of f at a point w with $|w| = r' < r$ may be explicitly expressed by an integral of $\operatorname{Im} f$ over a circle $|z| = r$ plus a term $\operatorname{Re} f(0)$. Hence, for instance,

$$M(f, r/2) \le \operatorname{const} A(f,r) + |f(0)|$$

so that, since $M(f,r)$ tends to infinity with r,

$$M(g, r/2) = O(\log M(f,r)).$$

In particular, if $M(f,r) = O(e^{r^m})$ for some $m > 0$, then $M(g,r) = O(r^m)$. But then g is a polynomial and assumes all values, which is a contradiction. If $\exp^{(n)}$ denotes the function exp iterated n times, the same argument and induction show that an entire function f which does not take two values cannot have a bound

$$|f(z)| \le \exp^n(O(|z|^m))$$

for any integer $n > 0$ and number $m > 0$. It follows that no entire function of reasonable growth can avoid two values. This is also the point where Borel stops in his lectures.

In the beginning of the century Schottky (1904) gave the first proof after Borel of Picard's theorem without using a modular function. It was followed by a flurry of papers by Landau, Hurwitz, Caratheodory and others. In the second edition (1929) of his book *Ergebnisse* Landau gives a number of properties of an analytic function in a disk which avoids the values 0 and 1. One of them, called the theorem by Schottky, says that a function f which is regular in the unit disk and does not assume the values 0 and 1 has a bound for $|z| < \theta < 1$ which only depends on θ and a bound of $|f(0)|$ away from 0 and infinity.

From this theorem, Picard's theorem can be deduced as follows. Assume that $F(z)$ is analytic for $0 < |z| < 1$, has an essential singularity at the origin and is never equal to 0 or 1. Put

$$F(e^t) = g(t)$$

so that $g(t)$ is defined when $\operatorname{Re} t < 0$ and is periodic with the period $2\pi i$. By Weierstrass's theorem there is a sequence of radii r_n tending to zero and points z_n

on the corresponding circles where $|F(z_n) - 2| < 1/2$. Putting $z = e^t$ we then have $|g(t_n) - 2| < 1/2$ where $\operatorname{Re} t_n = \log r_n$ tends to $-\infty$. Now the function

$$h(u) = g(t_n + 4\pi u)$$

is analytic when $|u| < 1$, it is never 0 or 1 and $|h(0) - 2| < 1/2$. Hence, by Schottky's theorem, h has an absolute bound when $|u| < 1/2$ and this suffices to cover an interval between t_n and $t_n + 2\pi i$. Hence the function $F(z)$ has a uniform bound on all circles $|z| = r_n$ and this is a contradiction.

The shortest proof of Schottky's theorem uses again the inverse $J(z)$ of the modular function $I(z)$ defined above. In fact, let $f(z)$ be a function analytic in the unit disk with a fixed $a = f(0)$, assume that f does not take the values 0 and 1 and consider the function

$$g(z) = J(f(z))$$

which maps the unit disk to the upper half-plane and so has the form

$$g(z) = \frac{ce^{i\theta} - \bar{c}z}{e^{i\theta} - z}, \quad \operatorname{Im} c > 0,$$

with

$$\operatorname{Im} g(z) = \frac{\operatorname{Im} c(1 - |z|^2)}{|1 - e^{i\theta}z|^2}.$$

Hence if $\operatorname{Im} g(z)$ tends to zero for some z in a closed disk $D : |z| \le b < 1$, then $\operatorname{Im} c$ tends to zero and hence $\operatorname{Im} g(z)$ tends to zero uniformly for all $z \in D$. Similarly, if $g(z)$ tends to infinity for some $z \in D$, then c tends to infinity and hence $g(z)$ tends to infinity uniformly for all $z \in D$.

Now consider a family F of functions f analytic in the open unit disk for which $a = f(0)$ stays in a compact set not containing 0 and 1. With z restricted to the disk $D : |z| \le b < 1$, suppose that $f(z)$ comes very close to $0, 1$ or is very large for some $f \in F$ and some $z \in D$. Then $\operatorname{Im} g(z)$ must come very close to the real axis or be very large and hence all of $g(D)$ has this property uniformly. But this contradicts the assumption about $f(0)$ and this proves Schottky's theorem. A modern proof where the topological content of the theorem is evident is available in Nevanlinna (1953).

Ahlfors's topological proof

Ahlfors's paper (1935b) which gave him a Fields medal is actually a topological proof of the essential part of Nevanlinna's theory. We shall now sketch the idea of the proof and how it can be used to prove Picard's theorem that a function $f(z)$ which is analytic outside a circle $|z| = r_0$ can avoid at most one value without being meromorphic at infinity.

Ahlfors's proof uses the Euler index, i.e. the number of corners minus the number of lines plus the number of triangles in a triangulation of a two-dimensional set. The Euler index is known to be a topological invariant. For a bounded set in the plane it equals $1 - q$ where q is the number of holes in the set.

The basis of the proof is a theorem by Hurwitz about covering maps $T : \bar{S} \to S$ of two-dimensional compact manifolds. If $T\bar{S}$ covers S N times, the theorem says that

$$\chi(\bar{S}) = N\chi(S) - \sum(\nu(P) - 1)$$

where χ is the Euler index, P runs through the points of S and $\nu(P)$ is the number of points of $\bar{S}$ over P. The proof[2] is immediate if we use all multiple points as corners in a triangulation of $\bar{S}$. It follows as a special case that

$$\chi(\bar{S}) \le N\chi(S).$$

Imagine now f as a map from the region $S_0 : R < |z| < \infty$ to a region S consisting of the complex plane C minus q points. We then have $\chi(S) = 1 - q$. Also, $\chi(S_0) = 0$, and this is also the Euler index of the image $\bar{S} = f(S_0)$. Hence, if we apply Hurwitz's theorem to this non-compact situation, we get

$$0 \le N(1 - q)$$

with some large N, perhaps infinity, and this means that $q \le 1$. This reasoning is of course complete nonsense, but in (1935b) Ahlfors got precisely this last inequality as a special case. Roughly speaking he arrived at this result by taking restrictions of f to ring-shaped regions $r_0 \le |z| \le r$ with a large r, by taking boundaries into account and by replacing N by a quotient of spherical areas. A fuller but not complete account of Ahlfors's arguments is given at the end of the chapter on Nevanlinna theory.

Bibliography

AHLFORS L.
    1935. a. *Über eine neue Methode in der Theorie der meromophen Funktionen*, Soc. Sci. Fenn. Comm. Phys.-Math. **8** (1935), no. 10, 1–14.
    1935. b. *Zur Theorie der Überlagerungsflächen*, Acta Math. **65** (1935), 157–194.
BOREL E.
    1897. *Sur les zéros des fonctions entières*, Acta Math. **20** (1957), 357–396.
    1904. *Leçons sur les fonctions entiéres*, Paris, 1900.
LANDAU E.
    1929. *Darstellung und Begründung einiger neuerer Ergebnisse der Funktionentheorie*, Berlin 1929.
NEVANLINNA R.
    1929. *Le théorème de Picard-Borel et la théorie des fonctions méromorphes*, Coll. Borel, Paris, 1929.
    1936. *Eindeutige Analytische Funktionen*, Springer, 1936, zw. Auflage, 1953.
PICARD E.
    1879. *Sur une propriété des fonctions entiéres*, Oeuvres I, p. 19.
    1880. *Mémoire des fonctions entières*, Oeuvres I, pp. 39–60.
SCHOTTKY F.
    1904. *Über den Picard'schen Sats und die Borel'schen Ungleichungen*, Sitzungsber. Der Kgl. Preussischen Akad. d. Wiss. Berlin Jahrgang 1904, pp. 1244–1262.

---

[2]The reader is advised to think through this argument.

CHAPTER 2

On Holmgren's Uniqueness Theorem

Introduction

The basic result about analytic solutions of partial differential equations with analytic coefficients is the Cauchy-Kovalevskaya theorem. Although it applies in great generality for non-linear equations, let us state it for the first order linear operator

$$P(x, D) = A_1(x)D_1 + \cdots + A_n(x)D_n + B(x), \quad D_k = \partial/\partial x_k,$$

where $A_1(x), \ldots, A_n(x)$ are $m \times m$ matrices which are analytic in the real variables $x = (x_1, \ldots, x_n)$. Let $P_1 = \sum A_k D_k$ be the principal part of P. A hypersurface $S : s(x) = 0$ is said to be characteristic for P at a point x if $P_1(x, s_x)$ is not invertible. By the Cauchy-Kovalevskaya theorem, the boundary problem

$$Pu = v, \quad u = w \quad \text{when} \quad s(x) = 0$$

has a local analytic solution u at non-characteristic points of S when P, w, s are analytic. In a sense, this result parametrizes the solutions of $P(x, D)u = v$ close to a non-characteristic point of S.

On the turn of the century there was a growing interest in classes of functions which are not analytic, only sufficiently differentiable in some sense, for instance those with continuously differentiable derivatives up to some order. One mathematician working in this area was the Swede Erik Holmgren, later professor at Uppsala university. He had the bright idea to ask himself what happens in the linear Cauchy-Kovalevskaya theorem when the solution u is not analytic and only sufficiently differentiable. The answer is given by Holmgren's uniquenss theorem (1901): if the solution exists, it is unique.

This lecture gives the simple proof of Holmgren's theorem followed by an example that it fails for non-analytic coefficients and a simple, recent example by Métivier that it fails also for non-linear systems with analytic coefficients. In a final section, the theorem is extended to non-analytic operators of the above form which are elliptic and almost commute with their adjoints. It illustrates the use of weight functions first introduced by Torsten Carleman (1939) in a uniqueness proof for first order systems in two variables. His result was extended to several variables by Calderón (1958). We shall follow the treatment in Hörmander (1985 III) which is close to Carleman's. Hörmander's complete presentation of Calderón's theorem (1985 IV) is beyond the scope of a simple lecture.

At present, Holmgren's uniqueness theorem is just a convenient heading for various developments connected with the original result (see Hörmander (1994)).

Proof of Holmgren's uniqueness theorem

By an analytic change of the variables x and a linear change of the unknown function u we may assume that the coefficients of $P(x, D)$ are analytic at the origin, that S is given by $x_1 = x_2^2 + \cdots + x_n^2$ and that $A_1(x)$ is the unit matrix E. It suffices to prove that u vanishes close to the origin when $Pu = 0$, $u = 0$ on S. Let (u, v) be the Euclidean scalar product and let

$$P'(x, D) = -\sum D_k A_k'(x) + B'$$

be the adjoint of P so that $(Pu, v) - (u, P'v) = \sum D_k(A_k u, v)$. If we integrate over a region $K = K(c) : c > x_1 > x_2^2 + \cdots + x_n^2$ with upper boundary K_+ where $c = x_1$ and put $x' = (x_2, \ldots, x_n)$, we get

$$\int_{K_+} (u, v) dx' = \int_K (u, P'v) dx.$$

Here we can let v be an analytic solution of $P'v = 0$ with data on K_+. Since these quantities can be given arbitrarily, u must vanish on every K_+. Varying the size and position of $K(c)$ shows that $u = 0$ close to the origin.

No uniqueness

For non-analytic coefficients Holmgren's uniqueness theorem is no longer true and several counterexamples were constructed in the 1950's. A general construction is presented in Hörmander (1983), a simple example of which is the following: there is a C^∞ function $a(t, x)$ which vanishes for $t \leq 0$ and whose support contains the origin such that the equation

$$\partial_t u + a \partial_x u = 0$$

has a solution $u = f(t, x) \neq 0$ which vanishes for $t \leq 0$. It was remarked by Métivier (1993) that this permits construction of a non-linear *analytic* system of equations

$$\partial_t u + v \partial_x u = 0, \quad \partial_t v + \partial_y v = 0$$

for which Cauchy's problem with data on $t = 0$ has two different solutions u, v which coincide for $t = 0$. It suffices to put either

$$u = g(y) f(t - y, x), \quad v = a(t - y, x)$$

where $g \in C^\infty$ is supported in $y \geq 0$ and vanishes otherwise, or

$$u = 0, \quad v = a(t - y, x).$$

The two solutions are smooth and different, but they are equal when $t = 0$. Hence Holmgren's theorem cannot be true for non-linear analytic systems. Métivier also has a couple of other counterexamples.

Uniqueness for non-analytic coefficients

Holmgren's result is much more difficult to prove for operators with non-analytic coefficients. The first proof in this case is due to Carleman (1939). He treated linear first order operators in two variables of the form

$$P(x, D) = D_1 u + A_2(x) D_2 + B(x), \quad D_k = \partial/i \partial x_k,$$

(since we shall use the Fourier transform later, we now use the imaginary gradient D). It is important that the square matrix $A_2(x)$ may be uniformly diagonalized so that, by a change of variables, Carleman could assume that $A_2(x)$ is already diagonalized. This implies in particular that

(1) $$[P(x, D), P^*(x, D)] = O(|u(x)||Du(x)|), \quad P^*(x, D) = \bar{P}'(x, D),$$

provided the coefficients are uniformly Lipschitz continuous. We shall assume (1) also in the general case where

$$P(x, D) = \sum_1^n A_k(x) D x_k + B(x)$$

and the coefficients are $m \times m$ Lipschitz continuous matrices. When the coefficients of $P(x, D) = P(D)$ are constant and (1) holds, the right side vanishes, so that $P(D)$ and $P^*(D)$ commute. When $n > 2$ it seems difficult to imitate Carleman by making preliminary changes of the coefficients in order to achieve (1).

Our proof of the uniqueness theorem below is parallel to the proof in Hörmander (1985 III) of a corresponding result in the scalar case where the analogue of (1) is automatic.

Outline of the theorem and a proof for constant coefficients

Let K be the region

$$K : c > x_1 > b(x_2^2 + \cdots + x_n^2), \quad 1 > b, c > 0.$$

We shall consider solutions u of the inequality

(*) $$|P(x, D)u| = O(|u|), \quad x \in K,$$

for small x_1. Here u with locally square integrable derivatives is assumed to vanish below the lower boundary $S : x_1 = b(x_2^2 + \cdots + x_n^2)$ of K. The coefficients $A_k(x), B(x)$ are supposed to be bounded and Lipschitz differentiable in $\bar{K}$. We shall find conditions under which such a u vanishes for small x_1. The main tool of the proof is the use of a function $h(x) = x_1 - x_1^2$ for small $x_1 > 0$ and a corresponding change of the unknown function,

$$v = e^{-h(x)} u, \quad u = e^{h(x)} v.$$

Then $D_j e^h v = e^h(D_j - ih_j)v$ where $h_j = \partial_j h$ and

$$\int_K e^{-2h} |P(x, D)u|^2 dx = \int_K |P(x, D - i\partial h)v|^2 dx.$$

Estimates referring to such norms were first called Carleman's estimates by Lars Hörmander.

Let us now note that the commutator

$$[D_j - ih_j(x), D_k + ih_k(x)]$$

vanishes unless $j = k = 1$, in which case it equals $h_{11}(x) = -2$. From this simple formula and (1) follows the main ingredient of our future proof, namely the identity

$$(2) \quad \int_K |P(x, D - \tau ih'(x))v(x)|^2 dx = \int_K |P^*(x, D + \tau ih'(x))v(x)|^2 dx$$
$$+ 2\tau \int_K |v(x)|^2 dx + \int_K O(|v(x)||Dv(x)|)dx$$

where $\tau > 0$. When P has constant coefficients, this holds without an error term and the desired uniqueness follows by inserting (*). In fact, then

$$2\tau \int_K e^{-2\tau h(x)} |u(x)|^2 dx \le \int e^{-2\tau h(x)} O(|u(x)|^2) dx.$$

Hence, letting τ tend to infinity, it follows that $u = 0$ in K.

Permitted simplifying assumptions

In the general case, one can try to find properties of $P(x, D)$ besides (1) which make the first integral on the right in (2) so positive for large τ that the kind of argument just given goes through. Before proceeding further we shall now state some permitted simplifications which influence neither the assumption (*) nor the conclusion of the theorem.

1) $B(x) = 0$.
2) The solution u has compact support close to the origin.
3) $P(x, D)$ is replaced by

$$P_\varepsilon(x, D) = \varepsilon P(\varepsilon x, D_{\varepsilon x}) \doteq \sum A_k(\varepsilon x) D_k, \quad \varepsilon > 0.$$

To see 2), replace u by κu where $\kappa(x) \in C_0^1$ equals 1 for small x. The effect of 3) is to make the error term of (1) small when P is replaced by P_ε. In the sequel we shall make tacit use of these assumptions.

Ellipticity assumption and the full proof

Besides (1) we now assume that $P(0, D)$ is elliptic at the origin, i.e. that

$$(1') \qquad\qquad |P(0, \xi)a|^2 \ge C|\xi|^2 |a|^2, \quad a \in C^n.$$

With $N = h'(0) = (1, 0, \ldots, 0)$, this means in particular that

$$|P^*(0, \xi + i\tau N)a|^2 + \tau^2|a|^2 \ge C(|\xi|^2 + \tau^2)|a|^2$$

with another constant $C > 0$. In fact, the left side vanishes only when $a = 0$ and it is homogeneous of order two in a and (ξ, τ). By a Fourier transformation, some easy estimates and a passage to P_ε^*, this shows that

$$\int (|P_\varepsilon^*(x, D + i\tau h'(x))v(x)|^2 + \tau^2|v|^2)dx \ge C \int (|Dv|^2 + \tau^2|v|^2)dx$$

with still another constant $C > 0$ when the support of u is close enough to the origin. We shall now use this inequality to minorize the right side of (2) and still get a term $\tau \int |v(x)|^2 dx$ left.

If $\tau > 1$, we may divide all terms of the inequality above except the first one by τ. After adding $\tau \int |v|^2 dx$ to both sides, the result is

$$(3) \quad \int |P_\varepsilon^*(x, D + i\tau h'(x))v(x)|^2 dx + 2\tau \int |v|^2 dx$$

$$\geq \tau \int |v|^2 dx + C \int (\tau |v|^2 + \tau^{-1}|Dv|^2) dx.$$

Here the last integral majorizes $2C \int |v||Dv| dx$ which in turn majorizes the last term of the second member of (2) applied to P_ε with $\varepsilon > 0$ small. In fact, the term in question then acquires a factor of ε. Hence, when ε is small enough, the right side of (3) suffices to balance this error term and leave, for instance, a term

$$\tau \int |v|^2 dx$$

so that

$$\tau \int e^{-2\tau h(x)} |u(x)|^2 dx \leq \int e^{-2\tau h(x)} |P_\varepsilon(x, D)|^2 dx$$

and the proof goes through as before.

Bibliography

CALDERÓN A. P.
  1958. *Uniqueness in the Cauchy Problem for partial differential equations*, Ann. of Math. **80** (1958), 16–36.
CARLEMAN T.
  1939. *Sur un problème d'unicité pour les systèmes d'équations aux dérivées partielles à deux variables réelles*, Mat. Astr. Fysik **26B** (1939), no. 17.
HOLMGREN E.
  1901. *Über Systeme von linearen partiellen Differentialgleichungen*, Öfversigt Vetensk. Akad. Handlingar **58** (1901), 91–105.
HÖRMANDER L.
  1983. *The Analysis of Linear Partial Differential Operators*, I, II, Grundlehren der Math. Wiss., vols. 256, 257, Springer, 1983.
  1985. *The Analysis of Linear Partial Differential Operators*, III, IV, Grundlehren der Math. Wiss., vols. 274, 275, Springer, 1985.
  1994. *Remarks on Holmgren's uniqueness theorem*, Ann. Fourier **43** (1994) (Coll. Malgrange), 1223–1251.
MÉTIVIER G.
  1993. *Counterexamples to Holmgren's uniqueness theorem for analytic non-linear Cauchy problems*, Inv. Math. **113** (1993), 217–222.

CHAPTER 3

The Phragmén-Lindelöf Principle

Introduction

The Phragmén-Lindelöf principle is an extremely useful extension of the maximum principle for analytic or harmonic functions. The first author, Edvard Phragmén (1863–1937), was a very gifted Swedish mathematician who succeeded Sonya Kovalevskaya as a professor at Stockholm University. He later entered the insurance business. The second author, Ernst Lindelöf (1870–1946), was the leading Finnish mathematician of his time. The principle itself was born in a mathematical atmosphere created by the father of Swedish mathematics, Gösta Mittag-Leffler (1846–1927), a very enterprising professor who founded the journal Acta Mathematica and became the leader of a group of mathematicians at Stockholm University who were very active around the turn of the century.

The Phragmén-Lindelöf principle (or PL principle for short) first appeared in a paper (1904) by Phragmén and got its final form in a joint paper (1908) by Phragmén and Lindelöf. Phragmén's inspiration came from a paper by his teacher Mittag-Leffler. The details of the story will be told below together with some applications and refinements of the PL principle. The most important one is due to the great Swedish mathematician Torsten Carleman (1892–1948), professor in Stockholm from 1924.

Phragmén's paper

Around 1900 summation of divergent series was a very popular subject. One of the main contributors was the French mathematician Émile Borel, author of a standard book (1901) on the subject. Another worker in the field was Mittag-Leffler. He devoted many years of work to a series of articles on the summation of Taylor series outside the circle of convergence.

One of Mittag-Leffler's methods, inspired by Borel, was based on an entire function $E_\alpha(z)$ defined by

$$E_\alpha(z) = \sum_0^\infty \frac{z^n}{(\alpha n)!}, \quad 0 < \alpha \le 1,$$

where $(\alpha n)! = \Gamma(1 + n\alpha)$. For $\alpha = 1$, this is the exponential function. Its properties become clear if we write it as an integral. By a known formula,

$$\frac{1}{\Gamma(1+\alpha)} = \frac{1}{2\pi i} \int_S e^u u^{-\alpha} du/u,$$

where S is a loop around zero, for instance given by $u = 1 + is - \varepsilon|s|$ with s real and $\varepsilon > 0$. The previous formula and a summation of the series for $E_\alpha(z)$ show that

$$E_\alpha(z) = \frac{1}{2\pi i} \int_S e^u \frac{u^\alpha}{u^\alpha - z} du/u$$

when $|z| < 1$, but then also when z is to the left of S. By a change of variables,

$$E_\alpha(z) = \frac{1}{2\pi i\alpha} \int_{S(\pi\alpha)} e^{u^{1/\alpha}} \frac{1}{u - z} du.$$

Here $S(\pi\alpha)$ bounds a sector around the real axis with the angle $\pi\alpha$. It is clear that $E_\alpha(z)$ is bounded and tends to zero at infinity in the complement of this sector. If z is in the sector, a residue $\alpha^{-1} \exp z^{1/\alpha}$ is added to the integral and represents the growth of E_α for large z in the sector. To sum up: $E_\alpha(z)$ tends to zero at infinity outside $S(\pi\alpha)$ and is large inside. The narrower the sector, the larger the growth of E_α.

Phragmén proved in (1904) that the behavior of $E_\alpha(z)$ is an example of a quite general phenomenon. In fact, let $f(z)$ be an entire analytic function such that

(1) $$\log|f(z)| \leq C + C|z|^a, \quad 0 < a < 1$$

in some proper sector S of the right half-plane with $f(z)$ bounded otherwise. In his proof Phragmén considered the integral

$$g(z) = \int_0^\infty f(zw)e^{-w} dw$$

which is absolutely convergent and defines an entire funtion $g(z)$. It is bounded when z is outside S, but the line of integration can move freely in the right half-plane, which means replacing z by $z' = ze^{i\theta}$ with $|\theta| < \pi/2$. Wherever z is in the sector S, z' can be moved outside the sector S by a suitable choice of θ. It follows that $g(z)$ is a bounded entire function and hence constant. But the coefficients of the Taylor series for f are just non-zero multiples of those of g so that f must be constant.

The paper with Lindelöf

Phragmén's paper was just the beginning and did not cut at the heart of the situation. The paper (1908) by Phragmén and Lindelöf had the right answer and the right proof. Suppose that $f(z)$ is analytic in the right half-plane and satisfies (1) there and that $|f(z)| \leq C$ on the imaginary axis. Let $1 > c > a$ and $\varepsilon > 0$ and consider the function

$$F_\varepsilon(z) = e^{-\varepsilon z^c} f(z).$$

One sees that $|F_\varepsilon(z)| \leq C$ on the imaginary axis and that $|F_\varepsilon(z)|$ tends to zero at infinity, whatever $\varepsilon > 0$. Hence there can be no regions where $|F_\varepsilon(z)| > C$ because these must be bounded and then the ordinary maximum principle applies. Hence $|f(z)| \leq C$ everywhere in the right half-plane.

By conformal maps this theorem carries over from the half-plane to any simply connected region Ω with a marked boundary point z_0. A rough statement of the PL principle is the following: if an analytic function $f(z)$ in Ω has the property that

$$\limsup |f(z)| \leq M$$

at all boundary points $z \neq z_0$ and $|f(z)|$ has less than a certain *critical growth* at z_0, then $|f(z)| \leq M$ everywhere in Ω.

What the critical growth is depends on the shape of Ω close to z_0. We have seen above that the critical growth at infinity in the right half-plane is $O(e^{|z|})$. A conformal map $z \to z^a$ shows that $O(e^{|z|^a})$ is the critical growth at infinity in a sector with opening angle π/a. The behavior of the function

$$e^{\cosh z}$$

in the strip $|y| < \pi/2$ shows that the corresponding critical growth at infinity is doubly exponential.

The details of the collaboration between Phragmén and Lindelöf are not known. For instance, who got the idea of connecting Phragmén's paper with the maximum principle? Finnish mathematicans argue privately that it was Lindelöf.

A digression on subharmonic functions

The PL principle concerns the absolute values of analytic functions. It can conveniently be transported to *subharmonic* functions, i.e. real functions defined in open plane regions, continuous from above and satisfying the maximum principle relative to harmonic functions in open regions. As is well known, it suffices to require this last property for disks and constant functions. Subharmonic functions obey the maximum principle: if $u(z)$ is subharmonic in an open, bounded region and $\limsup u$ at the boundary is at most M, then $u \leq M$ in the whole region. The main example of this notion is the following one:

The logarithm of the absolute value of an analytic function is subharmonic.

The critical growth of a subharmonic function in a sector with opening angle π/α is $|z|^\alpha$.

Carleman's sharper results

If we observe that $\log|f(z)|$ is harmonic outside the zeros of $f(z)$, we can sharpen the results above. In the theorem below the notation $\log^+|a|$ means $\log\max(|a|, 1)$.

THEOREM 1. *If $f(z)$ is analytic in the right half-plane and $|f(z)| \leq 1$ on the imaginary axis, then*

(2)
$$(\pi R)^{-1} \int_{-\pi/2}^{\pi/2} \log^+ |f(Re^{i\theta})| \cos\theta d\theta$$

is a non-decreasing function of R.

Remark. It follows that $|f(z)| \leq 1$ in the right half-plane if the right side of (2) is ≤ 0 for arbitrarily large R, in particular when $\log|f(z)| = o(|z|)$.

PROOF. The function v_R given by

$$\pi v_R(z) = \frac{\cos\theta}{r} - \frac{r\cos\theta}{R^2}, \quad r = |z|,$$

is harmonic in the right half-plane and vanishes on the imaginary axis and when $r = R$. Let us put $u(z) = \log^+|f(z)|$ and integrate $u\Delta v_R - v_R \Delta u$ over the region W where $u > 0$, $x > 0$, $|z| < R$. In the region $|z| < R$, the region W has a boundary

S intersecting $|z| = R$ in a set T. There are two boundary integrals whose sum vanishes,

$$\int_S v_R(s)u(s)_n ds, \quad \int_T u(s)v_R(s)_n ds$$

where the index n means the exterior derivative. The second integral is just the negative of (2). In the first one, $u(s)_n \geq 0$ and $v_R(s) > 0$ increase as R increases. In addition, the domain W of integration increases with R. Hence the theorem follows.

The theorem just proved is part of a general formula by Carleman which reads as follows:

$$\sum_{f(z)=0, 1<|z|<R} v_R(z) = U(R) + \frac{1}{2\pi} \int_1^R (y^{-2} - R^{-2}) \log|f(iy)f(-iy)| dy + O(1)$$

where $U(R)$ is given by (2). The proof is obtained from the preceding one by taking boundary terms on the imaginary axis into account and taking away a disk of radius one around the origin. The sum on the right runs over the zeros of f in the region between the two circles. Some consequences of this formula will be drawn below.

Some applications of the PL principle

The applications of the PL principle are not always direct. Here are some examples.

If u is subharmonic and $o(|z|)^2$ in the first quadrant and

$$u(x) \leq Cax, \quad u(y) \leq Cby,$$

then

$$u(z) \leq C(ax + by)$$

in the quadrant. For $u(z) - C(ax + by)$ is ≤ 0 on the boundary and has less than critical growth in the quadrant. More generally: in a sector with opening angle π/α there is interpolation between boundary values growing less than the critical growth. Interpolation can be used to prove

CARLSON'S THEOREM (1914). *If $u(z)$ is subharmonic in the right half-plane, if $u(z) \leq C + a|z|$ and $u(iy) \leq C - b|y|$, $b > 0$, on the imaginary axis, then $u = -\infty$.*

PROOF. We can take $C = 0$. As above we have $u(z) \leq ax - b|y|$ in the right half-plane. Hence also $u(z) + cx \leq (a + c)x - b|y|$ in the right half-plane. But the right side is ≤ 0 on the boundary of a proper sector of the half-plane. Hence it is ≤ 0 in the sector and so for instance $u(x) \leq -cx$ for all $c > 0$ so that $u = -\infty$.

Remark. Carleman's formula can be used to prove Carlson's theorem for analytic functions and many similar results. For instance, if a function is analytic in the right half-plane and $\log|f(z)|$ satisfies the requirements of the theorem, Carleman's formula gives a contradiction when $f \neq 0$. In fact, the left side of Carleman's formula is ≥ 0, the first term on the right is bounded and the second term tends to $-\infty$. Another of Carlson's results proved in the same way is the following: a function $f(z)$, analytic in the right half-plane and bounded on the imaginary axis with zeros in the integers must vanish when $\log|f(z)| = O(|z|)$. More general results, not given here, also present themselves.

Remark. Carlson's theorem is also a consequence of the following very important and useful result: If $u(z)$ is subharmonic in the right half-plane and $u(z) \leq Cx + D$ for some $C, D > 0$, then $u(z)$ has a least harmonic majorant in the right half-plane given by

$$\frac{1}{\pi} \int_{-\infty}^{\infty} \frac{xu(it)dt}{x^2 + (y - t)^2}.$$

It follows that the majorant equals $-\infty$ when, as in Carlson's case,

$$\int^{\infty} (u(it) + u(-it))dt/t^2 = -\infty.$$

Bernstein's theorem

If $f(z)$ is an entire function such that

$$\log|f(z)| = O(c|y|) + O(\log(1 + |z|))$$

and $|f(z)| \leq M$ on the real axis, then, by Theorem 1,

$$|f(z)| \leq Me^{c|y|}.$$

Such a function has the property that (Bernstein)

$$|f'(x)| \leq Mce.$$

In fact, it suffices to prove this when $x = 0$ and then, by Cauchy's inequalities,

$$|f'(0)| \leq \min Me^{cr}/r = Mce.$$

The minimal module

The theory of entire functions was a big subject in the beginning of this century, especially of those of finite order a, which means that

$$\log|f(z)| = O(|z|^{a+\varepsilon})$$

for all $\varepsilon > 0$ but no $\varepsilon < 0$. Entire functions of order < 1 possess simple canonical products

$$\text{const} \prod(1 - \frac{z}{z_k})$$

with rather sparse zeros z_k. The Swedish mathematician Wiman studied Mittag-Leffler's function

$$f(z) = \sum \frac{z^n}{\Gamma(an + 1)}, \quad 0 < a < 1,$$

which has order a and the growth

$$\log|f(z)| \sim \frac{\pi}{\sin \pi a} \operatorname{Re} z^a$$

outside small circles around the zeros. This means that the maximum modulus

$$M(r) = \max_{|z|=r} \log|f(z)|$$

and the minimum modulus

$$m(r) = \min_{|z|=r} \log|f(z)|$$

are of the same order outside the zeros and one should have

$$m(r) \sim M(r)^{\cos \pi a}$$

outside the zeros. Some 80 years ago, a lot of work went into the proof of this statement, which is particularly interesting for $0 < a < 1/2$ because then $\cos \pi a$ is positive. Since

$$\left| 1 - \left| \frac{z}{z_n} \right| \right| \leq 1 + \left| \frac{z}{z_n} \right|,$$

the worst situation for the minimum modulus relative to the maximum modulus occurs when all the zeros are negative. Hence we may assume that all the zeros are negative so that the minimum modulus is attained on the negative axis. If we consider $g(z) = f(\sqrt{z})$ instead, we are in the familiar right half-plane and have to consider the size of $g(z)$ in the right half-plane and its size on the imaginary axis. This connects the problem with the PL priciple, something which Wiman readily observed.

Some fine points of the PL principle in the right half-plane

It follows from the proof of the PL principle that a function $u(z)$ which is subharmonic in the right half-plane and has order $a < 1$ in the sense that

(3) $$u(z) = O(|z|^{a+\varepsilon}), \quad \varepsilon > 0,$$

cannot be much smaller, for instance

$$u(iy) = O(|y|^b), \quad 0 < b < a,$$

on the imaginary axis. For then the function

$$u - C \operatorname{Re} z^b - D = u - C \cos b\theta r^b - D, \quad z = re^{i\theta},$$

with C, D sufficiently large is ≤ 0 on the imaginary axis so that $u = O(|z|^b)$ by the PL principle. Hence, if (3) is the best estimate of u, it follows that

$$u(iy)/(1 + |y|^b)$$

is unbounded for all $b < a$. We shall use the PL principle to give a finer version of this result.

THEOREM 2. *Suppose that u is subharmonic of order $a < 1$ in the right half-plane and put*

$$M(r) = \max_{\theta} u(re^{i\theta}), \quad |\theta| \leq \frac{\pi}{2}.$$

Then the inequality

$$u(iy) > \cos \frac{\pi}{2}(a + \varepsilon) M(|y|)$$

has infinitely many solutions for every $\varepsilon > 0$. If u has order a but is of minimal type so that $\liminf M(r)/r^a = 0$, the same inequality holds for $\varepsilon = 0$.

Remark. The theorem is an old result by Wiman and Valiron. The proof which follows and the statement for $\varepsilon = 0$ are due to Beurling (see Kjellberg (1948)).

PROOF. We shall consider functions

$$v(z) = \delta \cos(a + \varepsilon)\theta r^{a+\varepsilon} + c(\delta), \quad z = re^{i\theta},$$

with $\delta > 0$ small and $c(\delta)$ large. They are harmonic in the left half-plane and $u(iy) \leq v(iy)$ for all y if $c(\delta)$ is large enough. We can then choose $c(\delta)$ so that $u(iy) = v(iy)$ for one $y = y_\delta$. By the PL principle, $u(z) \leq v(z)$ in the right half-plane so that

$$M(r) \leq \delta r^{a+\varepsilon} + c(\delta)$$

for all r. But now also

$$u(iy) = \delta \cos \frac{\pi}{2}(a + \varepsilon)r^{a+\varepsilon} + c(\delta)$$

for $y = y_\delta$. Hence, if $r = |y_\delta|$, we have

$$u(iy_\delta) - \cos \frac{\pi}{2}(a + \varepsilon)M(r) > (1 - \cos \frac{\pi}{2}(a + \varepsilon))c(\delta).$$

Since $|y_\delta|$ tends to infinity when δ tends to zero, the first part of the theorem follows. When $u(z)$ is of minimal type, the PL part of the same reasoning goes through with $\varepsilon = 0$.

Note. Returning to the entire functions $f(z)$ of order $< 1/2$, we have

$$m(r) > M(r)^{\cos \pi(a+\varepsilon)}$$

for an infinte sequence of radii for every $\varepsilon > 0$. When $f(z)$ is of minimal type, i.e. $\liminf M(r)/r^a = 0$, we can take $\varepsilon = 0$.

Bibliography

BOREL É.

    1896. *Fondements de la théorie des séries sommables*, J. de Math. Série 5, **2** (1896), 103–122.

    1899. *Mémoire sur les séries divergentes*, Annales de l'Ecole Normale Série 3, **16** (1899).

    1901. *Leçons sur les séries divergentes*, Paris, 1901.

CARLEMAN T.

    1923. *Über die Approximation analytischer Funktionen durch lineare Aggregate von Potenzen*, Arkiv Mat. Astr. Fysik **17** (1923).

CARLSON F.

    1914. *Sur une classe de séreis de Taylor*, Thesis, Uppsala, 1914.

KJELLBERG B.

    1948. *On certain integral and harmonic functions*, Thesis, Uppsala, 1948.

MITTAG-LEFFLER G.

    1900. *Sur la représentation analytique d'une branche uniforme d'une fonction monogène. Première note*, Acta Math. **23** (1900), 43–62.

    1901. a. *Seconde note*, Acta Math. **24** (1901), 183–204.

    1901. b. *Troisième note*, Acta Math. **24** (1901), 205–244.

    1902. *Quatrième note*, Acta Math. **26** (1902), 353–392.

    1905. *Cinquième note*, Acta Math. **29** (1905), 101–181.

    1920. *Sixième note*, Acta Math. **42** (1920), 285–308.

PHRAGMÉN E.

    1904. *Sur une extension d'un théorème classique de la théorie des fonctions*, Acta Math. **28** (1904), 351–368.

PHRAGMÉN E., LINDELÖF E.
    1908. *Sur une extension d'un principe classique de l'analyse et sur quelques propriétés de fonctions monogènes dans le voisinage d'un point singulier*, Acta Math. **31** (1908), 381–406.
WIMAN A.
    1905. *Sur une extension d'un théorème de Hadamard*, Arkiv Mat. Astr. Fysik **2** (1905).

CHAPTER 4

Nevanlinna Theory

Introduction

Ernst Lindelöf (1870–1946) is the father of Finnish mathematics.[1] He had his most productive period after the turn of the century, he wrote extensively on the theory of entire functions and he is the author of an influential book *Le calcul des résidues* (1905). He shares his best known result, the Phragmén-Lindelöf principle (1908), with the Swedish mathematician Edvard Phragmén. Almost single-handed Lindelöf created a powerful Finnish School in the field of analytic functions whose ramifications are present to this day.

The earliest and best known result of the Finnish School is Rolf Nevanlinna's theory of meromorphic functions and its improvements and clarifications by Lars Ahlfors. This theory, developed between 1920 and 1935, was inspired by Picard's theorem and contains it as a very special case. It was codified twice, first in the monograph (1929) and then in another one (1936) with a second edition (1953) which contains important improvements by Lars Ahlfors and other Finnish mathematicians.

The main novelty of Nevanlinna's theory was the introduction of a new growth function T for functions which are meromorphic in a disk or in a neighborhood of infinity.[2] In the main result, the second fundamental theorem, T serves as an essential bound of a sum of terms which express abnormal behavior of the meromorphic function at a number of points. Picard's theorem is a simple consequence. In his famous paper (1935) Lars Ahlfors gave a topological explanation of both the growth function and the second fundamental theorem. It is the object of this essay to trace the main developments of the theory without going into details. To a large extent this history is told in the monograph (1953) by Rolf Nevanlinna himself, but a short sketch may still be of some interest.

Jensen's formula

In 1899 a paper by the Danish telephone engineer Jensen carrying the weighty title *On a new and important theorem in function theory* appeared in Acta Mathematica in the form of a letter to its editor, Gösta Mittag-Leffler. At first Jensen proves that

$$\frac{1}{2\pi} \int_0^{2\pi} \log \left|1 - \frac{re^{i\theta}}{a}\right| d\theta$$

equals zero when $|r| < |a|$ and $\log r/|a|$ when $r > |a|$. If $f(z)$ is an analytic function in the disk $C : |z| \leq r$ with the zeros $a_1, \ldots, a_n \neq 0$ inside and none at the boundary,

---

[1] See Elfving (1981).

[2] We shall only deal with the second case.

this fact is used to prove Jensen's formula

$$\frac{1}{2\pi} \int_0^{2\pi} \log|f(re^{i\theta})|d\theta = \log|f(0)| + \log\frac{r^n}{|a_1 \cdots a_n|}$$

which gives a connection between the number of zeros of f inside C and the mean growth of $\log|f|$ at the boundary. When f has the poles $b_1, \ldots, b_m$ in C, the term

$$-\log\frac{r^m}{|b_1 \cdots b_m|}$$

should be added to the right side.

The fundamental theorem of algebra is an immediate consequence of Jensen's formula. In fact, $m \log r$ has to be the order of magnitude for both sides when f is a polynomial of degree m. The importance of Jensen's formula is that it provides a similar connection for entire functions and even meromorphic functions.

The simplest way to prove Jensen's formula is to differentiate the mean value

$$M(f,r) = \frac{1}{2\pi} \int \log|f(e^{i\theta})|d\theta$$

with respect to r when $f \neq 0$ for $|z| = r$. Then $\log f(z) = \log|f(z)| + i\arg f(z)$ is an analytic function of $\log z = \log|z| + i\arg z$ and so

$$(\frac{\partial}{\partial \log r} + i\partial_\theta) \log f(re^{i\theta}) = 0,$$

in particular

$$\partial_r \log|f(z)| = \frac{1}{r}\partial_\theta \arg f(z), \quad z = re^{i\theta},$$

and it follows that

$$dM(f,r) = \frac{dr}{2\pi r}\mathrm{var}_{|z|=r} \arg f(z).$$

This is Jensen's formula in differential form. Let us now integrate the two sides from some fixed $r_0 \geq 0$ to a large number r. If $\Omega(r)$ is the corresponding annulus, if $n(f,r)$ is the number of zeros of f in $\Omega(r)$ and $n(1/f,r)$ the number of poles and there are no poles or zeros at the boundary, the result is that

(1) $$M(f,r) - M(f,r_0) = N(f,r) - N(1/f,r)$$

where

$$N(f,r) = \int_{r_0}^r n(f,r)dr/r.$$

For $r_0 = 0$ this is Jensen's formula. Note that since f and $f - a$ have the same poles, the last term of the right side of f does not change if we replace f by $f - a$ where a is any complex number. This remark indicates that the formula (1) may not be the end.[3]

---

[3]Our $N(1/f,r)$ is Nevanlinna's $N(f,r)$. The reader may think of the capital N as indicating both 'number of' and 'nul'.

Nevanlinna's first proofs of his theory

The origin of Nevanlinna's theory is probably the paper (1924) where it is proved that a meromorphic function in the unit disk is a quotient of two bounded analytic functions if and only if the function

$$\frac{1}{2\pi}\int_0^{2\pi} \log^+|f(re^{i\theta}|d\theta + \int_{r_0}^r p(t)dt/t$$

is bounded as $r \to 1$. Here $\log^+ a$ is the positive part of $\log a$ and $p(t)$ is the number of poles in the annulus $r_0 < |z| < r < 1$, the choice of r_0 being of no importance.

This result indicates that it would be useful to rewrite Jensen's formula by splitting

$$\log|f(z)| = \log^+|f(z)| - \log^+ 1/|f(z)|$$

into a difference of non-negative numbers and splitting $M(f, r)$ into $m(f, r) - m(1/f, r)$ where

$$m(f, r) = \frac{1}{2\pi}\int \log^+|f(e^{i\theta})|d\theta$$

is the mean value of the positive part of $\log|f|$ over the circle $|z| = r$.[4] This mean value will be coupled with the number of poles of f when $r_0 < |z| < r$. The formula (1) now reads

(2) $$m(1/f, r) + N(f, r) = m(f, r) + N(1/f, r) + O(1)$$

where $O(1)$ stands for the contributions from the boundary $|z| = r_0$. In this formula $dN(f, r) = n(f, r)dr/r$ where $n(f, r)$ means the number of zeros of f when $r_0 < |z| < r$. This inventive rewrite of Jensen's formula has the following remarkable consequence:

FIRST FUNDAMENTAL THEOREM. *For a function $f \neq 0$ which is meromorphic in a neighborhood of infinity, the function*

(3) $$T_a(f, r) = m(1/(f - a), r) + N(f - a, r),$$

interpreted as

$$T(f, r) = m(f, r) + N(1/f, r)$$

when $a = \infty$, is independent of the number a modulo $O(1)$.

Note. Since the theory is only concerned with growth at infinity, this result makes it possible to define a characteristic function class $T(f, r)$ equal to any $T_a(f, r)$ modulo terms $O(1)$. In this sense $T(f, r) = T(1/f, r)$.

PROOF. If we apply (2) to the function $f - a$, the number of poles remain unchanged and since $\log^+|b - a| \leq \log^+(|b| + |a|) \leq \log^+|b| + |a|$ or, better, $\leq \log^+|b| + \log^+|a| + \log 2$, the theorem follows.

We can look at $T(f, r)$ as measuring the affinity of f to any number a in $\Omega(r)$. The affinity is large when the function $f(z) - a$ has many zeros in the region or when $f(z) - a$ is small for $|z| = r$. If $f = e^z$, it is not difficult to see that the first

---

[4]When $m(1/f, r)$ and $m(f, r)$ are interchanged, the characteristic function $T(f)$ becomes $m(f, r) + N(f, r)$. This form is the one used in the next section on the Shimizu-Ahlfors version.

term r/π of $T_a(f,r)$ dominates when $a = 0, \infty$ while the second term dominates in all other cases where $f(z) - a$ has plenty of zeros.

Nevanlinna's second fundamental theorem is closely connected with Picard's great theorem.

SECOND FUNDAMENTAL THEOREM. *With assumptions as in the previous theorem about f, let $a_1, \ldots, a_p$ be separate complex numbers. Then, with r tending to infinity,*

$$(4) \qquad \sum_1^p m(1/(f - a_k), r) \le 2T(f,r) - N_1(f,r) + A(r)$$

where $N_1(f,r,0) \ge 0$ and $A(r) = O(\log T(f,r))$ except for a set on which every power of r has bounded variation.

If we introduce the *defects*

$$(5) \qquad \delta(f, a_k) = \liminf \frac{m(1/(f - a_k), r)}{T(f,r)} = \limsup \frac{1 - N(f - a_k, r)}{T(f,r)},$$

it follows that

$$\sum \delta(f, a_k) \le 2.$$

In particular, f can avoid at most two values in every neighborhood of infinity, or even more, $N(f - a_k, r)/T(f,r)$ can tend to zero for at most two values a_k.

All the proofs of this theorem use some not too transparent juggling with the characteristics of f and f'. This can only be sketched here. The function

$$F(z) = \sum \frac{f(z)}{f(z) - a_k}$$

and the mean value $m(F, r)$ over large $|F|$ play a decisive part. Since the main contributions come from places where $f(z)$ is close to some a_k, it is not surprising that

$$m(F, r) \ge \sum m(1/(f - a_k), r) + O(1).$$

To get an upper bound, F is rewritten as

$$(6) \qquad F = \frac{1}{f} \frac{f}{f'} \sum \frac{f'}{f - a_k}$$

with the idea that

$$(7) \qquad m(f'/f, r) \quad \text{is often small relative to} \quad \log T(f,r).$$

The computation runs as follows. By (6),

$$(8) \qquad m(F, r) \le m(1/f, r) + m(f/f', r) + O(m(f'/f, r)) + O(1).$$

The first term on the right is rewritten according to Theorem 1,

$$m(1/f, r) = T(f,r) - N(f,r) + O(1),$$

and so is the second term,

$$m(f/f', r) = m(f'/f, r) + N(f'/f, r) - N(f'/f, r).$$

Since $N(f_1 f_2, r) \leq N(f_1, r) + N(f_2, r)$, some computation proves that the sum of the second and third terms on the right of (8) is

$$T(f,r) - m(f,r) - N_1(f,r) + O(1),$$

where

$$N_1(r) = 2N(1/f, r) - N(1/f', r) + N(f', r)$$

denotes the number of double poles and zeros in $\Omega(r)$. Comparing the lower and upper bounds of $m(F,r)$ gives the desired formula (4). The proof of (7) is the difficult point of Nevanlinna's theory. A short proof is available in Nevanlinna (1953), pp. 254–255. An analogous statement for estimates of a positive function $f(r)$ in terms of its integral

$$g(r) = \int_1^r f(r) dr$$

runs as follows: $f(r) = O(r^k g(r)^a)$, $a > 1$ for a set A of numbers r such that

$$(9) \qquad \int_A r^k dr < \infty.$$

In fact, on A we have

$$(10) \qquad \int_A r^k dr = \int r^k g'(r) dr / f(r) \leq \int_A g'(r) dr / g(r)^a < \infty.$$

The Shimizu-Ahlfors version of the theory

Ahlfors (1906–1996) was fourteen years younger than Rolf Nevanlinna (1892–1995). Very soon after Ahlfors had entered the university he was accepted by Rolf Nevanlinna as a student and companion when visiting the mathematical centers of Europe. The results were astonishing. Ahlfors proved a number of important theorems and in (1935b) he gave a topological explanation of Nevanlinna's second fundamental theorem which earned him one of the two first Fields medals in 1936. Here we shall sketch the paper (1935a) which uses stereographic projection to prove Nevanlinna's theory and to give an important geometric interpretation of the characteristic function T.

Stereographic projection

The stereographic projection maps points in the complex plane to points on a unit sphere S which touches the complex plane at the origin. The image $s(a)$ of a number a is the point where a line through a and the north pole of S meets S. The distance between the images of two complex points a, b turns out to be

$$[a, b] = \frac{|a - b|}{\sqrt{1 + |a|^2}\sqrt{1 + |b|^2}}.$$

This proves that stereographic projection is conformal,

$$|ds(a)| = \frac{|da|}{1 + |a|^2},$$

and

$$d\omega(s(a)) = d\sigma(a)/(1 + |a|^2)^2,$$

where $d\omega$ is the surface element of the sphere and $d\sigma(a)$ the euclidean surface element in the plane. In particular,

$$(11) \qquad |f'(z)|^2 d\omega(f(z))$$

is the surface element of the spherical image of the map $z \to f(z)$ when f is meromorphic.

A geometric interpretation of the characteristic function

Suppose that $f(z) \neq 0$ is a meromorphic function when $|z| \geq r_0$. In (1935a) Ahlfors introduces a modified characteristic function of $f(z)$. The first step is to replace the earlier mean deviation of f from some value a on the circle $|z| = r$ by

$$m(f, a, r) = m(r, a) = \frac{1}{2\pi} \int_0^{2\pi} \log \frac{1}{[f(z), a]} d\theta, \quad z = re^{i\theta},$$

where a now may be infinity. Here it is convenient to differentiate a difference $m(r, a) - m(r, b)$ with respect to a. The result is

$$m'(r, a) - m'(r, b) = \int_0^{2\pi} \partial_r \log \left| \frac{f - b}{f - a} \right| d\theta$$

where the right side equals

$$\frac{1}{2\pi r} \int_0^{2\pi} \partial_\theta \arg \frac{f - b}{f - a} = \frac{n(r, b) - n(r, a)}{r}$$

with $n(r, a)$ the number of a-places of f in the annulus $r_0 < |z| < r$. Hence

$$m'(r, a) + \frac{n(r, a)}{r} = m'(r, b) + \frac{n(r, b)}{r}$$

so that by an integration

$$(12) \quad T(r) = m(r, a) + N(r, a) = m(r, b) + N(r, b), \quad N(r, a) = \int^r n(r, a) dr/r$$

for all a and b where $T(r_0) = 0$ may be achieved by a suitable choice of the constant of integration.

If we now integrate (12) over the sphere S with a normalization of the surface such that $\int d\omega(a) = 1$, the average of $[f(z), a]$ is some constant and we get

$$T(r) = \int_{r_0}^r N(r, a) d\omega(a).$$

Here (12) shows that the main term equals

$$(13) \qquad T(r) = \int_{r_0}^r S(r) dr/r$$

where $S(r) = \int n(r, a) d\omega(a)$. Changing variables to $a = f(z)$ the result is that

$$(14) \qquad S(r) = \int n(r, a) d\omega(a) = \int_{r_0 \leq |z| < r} |f'(z)|^2 d\omega(f(z))$$

is the spherical area of the image of the annulus $r_0 < |z| < r$ under the map $z \to f(z)$. This is the geometrical interpretation of the characteristic function, taken in the sense of (12). The formula (13) shows that $T(r)$ is a convex, increasing function of $\log r$.

The stage is now set for the new proof of the second fundamental theorem in Ahlfors (1935b). But this proof will not be sketched here. Instead we shall pass to his topological proof in (1935b) now with more detail than at the end of the paper on Picard's great theorem.

The topological proof of the second fundamental theorem

A function $f(z)$ which is meromorphic in a neighborhood of infinity where it has an essentially singular point maps the annulus $\Omega(r) : r_0 \leq |z| \leq r$ with r large to the sphere S of stereographic projection. By Weierstrass's theorem the image $F(r) = f(\Omega(r))$ is dense in S for all large r. Let $Q = (a_1, \ldots, a_q)$ be a finite number of separate points of S. To see how well the image $F(r)$ can cover $S \backslash Q$ we shall with Ahlfors subtract from $F(r)$ the set $P(r)$ of those interior points which are projected to some point in Q and apply Hurwitz's theorem to the map f from $\Omega(r) \backslash Q$ to $F(r) \backslash P(r)$. With χ denoting the Euler index, this gives the inequality

(15) $$\chi(F(r) \backslash P(r)) \leq N(r) \chi(\Omega(r) \backslash Q)$$

for some $N(r)$ which is the number of times that $F(r)$ covers S. Here, since $\chi(\Omega(r)) = 0$ and a hole in a surface subtracts one from its Euler characteristic, we have

(16) $$\chi(F(r) \backslash P(r)) = -\sum_{1}^{q} n(a_k, r),$$

where $n(a_k, r)$ is the number of times the point a_k is covered by open parts of $F(r)$. On the other hand we may expect that

(17) $$\chi(F(r) \backslash P(r)) \leq N(r) \chi(S \backslash Q)$$

since $F(r) \backslash P(r)$ covers the set $S \backslash Q$ $N(r)$ times. Here $\chi(S \backslash Q)$ equals $2 - q$, so that, dividing by $N(r)$ and passing to the limit, it is reasonable to expect that

$$\sum_{k} \liminf(1 - \frac{n(a_k, r)}{N(r)}) \leq 2.$$

Now since $S \backslash Q$ and $F(r) \backslash P(r)$ are not compact and $N(r)$ is not well determined, the inequality (17) is senseless but may still contain some information. This is verified in Ahlfors's paper (1935b). There it is shown that

(18) $$\chi(F(r) \backslash P(r)) \leq S(r) \chi(S \backslash Q) + \text{const } L(r) + \text{const},$$

where $S(r)$ is the quotient of areas of the spherical images of $\Omega(r)$ and $F(r)$ and $L(r)$ is the quotient of the corresponding lengths of the circle $|z| = r$ and the boundary of $F(r)$. Also,

$$\liminf L(r)/S(r) = 0.$$

Insertion of the Euler indices (17) and passage to the limit show, finally, that

$$\sum_{k} \liminf(1 - \frac{n(a_k, r)}{S(r)}) \leq 2.$$

Here the terms of the left side are analogous to the Nevanlinna defects $\delta(a_k)$ of the chosen points a_k (see (5)). The difference is that the numbers $n(a_k, r)$ and the area $S(r)$ now have replaced their integrals with respect to dr/r, namely the number function $N(a_k, r)$, and the characteristic function $T(r)$. The precise connections between the two variants are sorted out in Ahlfors's famous paper (1935b) but they are out of bounds for this short review. But it should be remarked that Ahlfors's topological result is valid for more general maps than those given by meromorphic functions, or, in other words, maps which are conformal in the spherical metric. Essential parts of the theory extend even for dimensions > 2 to quasi-regular maps for which the dilation varies within uniform bounds (see Rickman (1993)).

Bibliography

AHLFORS L.

   1935. a. *Über eine Methode in der Theorie der meromophen Funktionen*, Coll. Papers I, Birkhäuser, 1982.

   1935. b. *Zur Theorie der Überlagerungsflächen*, Coll. Papers I, Birkhäuser, 1982.

ELFVING G.

   1981. *The history of mathematics in Finland 1828–1918*, Helsinki, 1981.

NEVANLINNA ROLF.

   1924. *Über eine Klasse von meromorphen Funkionen*, Math. Ann. **92** (1924).

   1929. *Le théorème de Picard-Borel et la théorie des fonctions méromorphes*, Coll. Borel, Paris, 1929.

   1936. *Eindeutige analytische Funktionen*, Springer, 1936, zweite Auflage, 1953.

RICKMAN S.

   1993. *Quasi-regular Mappings*, Ergebnisse der Mathematik und ihre Grenzgebiete, 3. Folge, Band 26, Springer-Verlag, 1993.

CHAPTER 5

The Riesz-Thorin Interpolation Theorem

Introduction

The best known theorem by Marcel Riesz, my teacher, is the result that a function and its conjugate are in the same L^p class when $1 < p < \infty$. The proof combines some simple remarks with a new interpolation theorem later improved by Riesz's student Olof Thorin who found a completely different proof. The Riesz-Thorin theorem is now one of the gems of analysis. We shall follow the historical development. It consists of two papers (1927a,b) by Marcel Riesz and a paper (1938) and a thesis (1948) by Thorin. The further developments can be seen in Zygmund [1] and in a lot of papers on interpolation in Hilbert spaces.

Conjugate functions

Two harmonic functions $u(z), v(z)$ are said to be conjugate if $f(z) = u(z) + iv(z)$ is analytic. We shall consider conjugate functions on the real axis R coming from an analytic function $f(z)$ in the upper half-plane which vanishes at infinity. From Cauchy's theorem,

$$f(x + iy) = \frac{1}{2\pi i} \int_R \frac{f(t)dt}{t - x - iy},$$

and the fact that

$$\lim_{y \to 0}(t - x - iy)^{-1} = \mathrm{pv}\,\frac{1}{t - x} + \pi i \delta(t - x)$$

we see that $v = Tu$, where

$$Tu(x) = \frac{1}{\pi} \int \mathrm{pv}\,\frac{1}{x - t} u(t)dt$$

is called the Hilbert transform. It has the property that $T^2 = -1$. Hilbert proved that it is an isometry of L^2. The first property follows since $u + iv = i(v - iu)$. The second one is a simple consequence of the fact that

(1)
$$\int_R (u(x + iy) + iv(x + iy))^{2m} dx = 0$$

for functions $f = u + iv$ which are sufficiently small at infinity. Already the case $m = 2$ together with a passage to the limit, y tending to zero, shows that

$$\int u^2(x)dx = \int v^2(x)dx, \quad \int u(x)v(x)dx = 0.$$

Riesz's theorem about conjugate functions, first proof

The theorem says that $u \in L^p$ implies that $v \in L^p$ when $1 < p < \infty$ with a corresponding inequality

(2) $$\|v\|_p \leq C_p \|u\|_p.$$

Riesz's simple idea in the paper (1927a) is to use (1) to prove (2) for $p = 2, 4, 6, \ldots$ and to proceed from there. Simply consider that

$$\mathrm{Re}(u + iv)^{2m} = \sum \binom{2m}{2j}(-1)^{m-j}u^{2j}v^{2m-2j}.$$

and estimate

$$|u^{2j}v^{2m-2j}| \leq \varepsilon u^{2m} + c_\varepsilon v^{2m}$$

where ε is as small as we please. If we combine this with (1), we get $\|v\|_{2m} \leq$ const $\|u\|_{2m}$ and (2) follows for $p = 2, 4, 6, \ldots$. So far, this is a bit formal, but if we take $u \in C_0^\infty$, then $v = O(1/|x|)$ for large x and hence $v \in L^p$ for $p > 1$ so that also $f(x + iy) \in L^p$ for all $y > 0$. Hence, by a passage to the limit, (2) holds in the strict sense for $p = 2, 4, \ldots$. But it is clear that T is selfadjoint, $(Tf, g) = (f, Tg)$ for reasonable f, g and hence, by Hölder's inequality

$$\|Tu\|_{p'} = \sup |(Tu, g)|/\|g\|_p = \sup |(u, Tg)|/\|g\| \leq C_p \|u\|_{p'}$$

where $1/p + 1/p' = 1$ and (2) is supposed to hold for p. Hence (2) holds also when $p = 2m/(2m - 1)$ for $m = 1, 2, \ldots$. To prove the inequality (2) for the missing values of p, Riesz repeated the proof above with a broken power $f(z)^p$ of the analytic function $f(z) = u(z) + iv(z)$. He could do that when $u \geq 0$, but ran into troubles otherwise. They were eventually overcome, but the whole thing took another turn in his paper (1927b) where the Riesz convexity theorem is used to fill in the gaps. Therefore we shall turn to this paper and its extension by Thorin.

Remark. The computations above hold also in the unit disk for analytic functions vanishing at the origin. Their real and imaginary parts on the unit circle are then conjugate functions, simultaneously in $L^p(0, 2\pi)$ when $1 < p < \infty$.

Riesz's convexity theorem

Let

$$A(x, y) = \sum_{1 \leq j \leq n, 1 \leq k \leq m} a_{jk}x_j y_k$$

be a complex bilinear form and put

$$\|x\|_p = \left(\sum_1^n |x_k|^p\right)^{1/p}, \quad 1 \leq p \leq \infty$$

where $\|x\|_\infty$ is defined as $\max |x_k|$. Analogously for $\|y\|_q$ with the same restrictions for q.

RIESZ'S CONVEXITY THEOREM. *The function*

$$M(\alpha, \beta) = \inf |A(x, y)|/\|x\|_p\|y\|_q, \quad \alpha = 1/p, \ \beta = 1/q$$

is logarithmically convex in the triangle

(3) $$1 \geq \alpha, \beta \geq 0, \quad \alpha + \beta \geq 1.$$

Note. This means of course that

$$\log M(t\alpha_1 + (1-t)\alpha_2, t\beta_1 + (1-t)\beta_2) \leq t \log M(\alpha_1, \beta_1) + (1-t) \log M(\alpha_2, \beta_2)$$

when the two pairs (α_1, β_1) and (α_2, β_2) belong to the triangle (3).

PROOF. If we define the vectors X and Y by the equalities

$$A(x, y) = \sum x_k Y_k = \sum y_j X_j,$$

Hölder's inequality shows that

$$M(\alpha, \beta) \leq |A(x, y)|/\|x\|_p \|y\|_q \leq \|x\|_p \|Y\|_{p'}/\|x\|_p \|y\|_q$$

and hence

$$\|Y\|_{p'} \leq M(\alpha, \beta)\|y\|_q$$

where p' is dual to p in the sense that $1/p + 1/p' = 1$.

Using Lagrangian multipliers, one can verify that

$$M|x_k|^{p-1} = |Y_k|, \quad M|y_k|^{q-1} = |X_k|$$

at the point x, y where the minimum is obtained. Hence

$$M\||x|^{p-1}\|_{p_1'} \leq \|Y\|_{p_1'} \leq M_1\|y\|_{q_1}$$

and

$$M\||y|^{q-1}\|_{q_2'} \leq \|X\|_{q_2'} \leq M_2\|x\|_{p_2}$$

where M_1 refers to p_1, q_1 and M_2 to p_2, q_2. Raising the first of these formulas to the power t and the second to the power $1 - t$ and multiplying them gives Riesz the vehicle with which he proves his convexity theorem. The computations are elementary but tedious and will not be given here.

Passing through functions with a finite number of values, Riesz's convexity theorem provides a new proof of the theorem on conjugate functions and other inequalities. This will be done later in this lecture after Thorin's extension of Riesz's convexity theorem has been presented.

Riesz was troubled by the fact that his proof did not extend convexity to the entire unit square $1 \geq \alpha, \beta \geq 0$ and at the end of his paper he even conjectured that this was not possible. But ten years later the extension was done by Riesz's student Olof Thorin (1938). Shortly after Thorin had published his proof, he went into the insurance business. In his thesis (1948) he enlarged on the subject and included a number of applications.

Thorin's extension

Thorin's idea was to transform the quotient

$$C = A(x, y)/\|x\|_p \|y\|_q$$

into an analytic function of $\alpha = 1/p, \beta = 1/q$ and employ a maximum principle. To do this, write

$$x_j = c_j^\alpha e^{i\theta_j}, \quad y_k = d_k^\beta e^{i\varphi_j}$$

where $c_j, d_k \geq 0$. We are free to choose these quantities so that $\sum c_j = \sum d_k = 1$. Then

$$C = \sum a_{jk} e^{i\theta_j + i\varphi_k} c_j^\alpha d_k^\beta.$$

With α, β replaced by $\alpha_1 + s\beta_1, \alpha_2 + s\beta_2$, C appears as an analytic function $f(s)$ which, if α, β are in the square, is analytic when $0 \leq \operatorname{Re} s \leq 1$ and bounded. Hence, by the Phragmén-Lindelöf principle,

$$\sup_{\operatorname{Re} s = t} \log |f(s)|$$

is a convex, bounded function. Taking the supremum over all arguments θ_j, φ_k and amplitudes c_j, d_k does not change the convexity and hence Riesz's convexity theorem is extended to the square.

Applications

The restriction of the theorem above to finite bilinear forms is not essential. Let $M = (x, y, \dots)$ be a measure space with a completely additive measure $d\mu$ and let $L^p(M)$, $1 \leq p \leq \infty$, be the space of measurable complex functions f from M such that

$$\|f\|_p = (\int_M |f(x)|^p d\mu(x))^{1/p} < \infty$$

modulo those functions for which the left side vanishes. When $p = \infty$, the left side should be the essential supremum of $|f|$ defined in the usual way. Let $B(f, g)$ be a bilinear functional on the product of two measure spaces M_1, M_2, defined when f and g have a finite number of values. Then, by reduction to finite bilinear forms,

$$\sup |B(f, g)| / \|f_p\| \|g\|_q,$$

where f is defined on M_1 and g on M_2, is a convex function of $1/p, 1/q$ in the unit square. In fact, this is true when f and g both have a fixed but arbitrary number of values. Such spaces are dense in $L^p(M_1)$ and $L^q(M_2)$ and hence, if the supremum is finite for two sets of values p_1, q_1 and p_2, q_2, it is finite for all p, q such that $1/p, 1/q$ lies on a straight line between $1/p_1, 1/q_1$ and $1/p_2, 1/q_2$.

The simplest application is Hölder's inquality,

$$|\int_M f(x)g(x)d\mu(x)| \leq \|f\|_p \|g\|_{p'}$$

when f and g are defined on the same measure space. The inequality holds trivially when $p = 1, \infty$. Hence it holds in general.

In order to apply the theorem to conjugate functions, use the bilinear form

$$B(f, g) = \int f(x) Tg(x) dx.$$

By Riesz's simple computations with even powers

$$|B(f, g)| \leq C_p \|f\|_{p'} \|g\|_p$$

when $p = 2n$ or $p' = 2n$ for all integers $n > 0$. Hence, by Hölder's inequality (or rather its converse), it holds in general and hence

$$\|Tg\|_p \leq C_p \|g\|_p$$

with $C_p < \infty$ for all $1 < p < \infty$.

I leave it to the reader to prove in the same way the Hausdorff-Young inequality

$$\|\hat{f}\|_p \leq C_{pq} \|f\|_q$$

where $1/p + 1/q = 1$, $q \geq 2$ and

$$\hat{f}(\xi) = \int_R e^{-i(x,\xi)} f(x) dx$$

is the Fourier transform of f.

Bibliography

THORIN O.

1938. *An extension of a convexity theorem due to M. Riesz*, Fys. Sällsk. Förh. **8** (1938), no. 14.

1948. *Convexity theorems generalizing those of M. Riesz and Hadamard*, Ak. Avh. Lund, 1948.

RIESZ M.

1927. a. *Sur les fonctions conjuguées*, Math. Z. **27** (1927), 218–244.

1927. b. *Sur les maxima des fonctions bilinéaires et sur les fonctionelles linéaires*, Acta Math. **49** (1927), 465–497.

CHAPTER 6

The Mathematics of
Wiener's Tauberian Theorem

Introduction

Wiener's long paper (1932) is one of the true classics of pre-war mathematics. The main result, one of the first results in the paper, says that the closed linear hull of translates of a function

$$f(x) \in L^1 = L^1(R)$$

is the whole space L^1 if and only if its Fourier transform

$$F(x) = \mathcal{F}f(x) = \int_R e^{-itx} f(t) dt$$

never vanishes. This basic result has been called Wiener's Tauberian theorem, an unnecessarily complicated name whose origin will be explained below.

Note that the closed linear hull in question contains all convolutions

$$f * g(x) = \int f(x - y)g(y) dy.$$

In fact this holds when g is a continuous function with compact support and we have the inequality

$$\|f * g\| \le \|f\| \, \|g\|$$

where

$$\|f\| = \int_R |f(x)| dx.$$

About ten years after Wiener's paper, Gelfand's papers on normed rings or, with a later term, Banach algebras, started the field of Abstract Harmonic Analysis, which was intensely studied for almost twenty years, especially in the United States. One of the driving forces of this theory was the desire to extend Wiener's theorem to the new abstract landscape. The book (1962) by Walter Rudin summarizes most of this research.

The name Tauberian theorem has a long and winding history. In such a theorem one knows the asymptotic behavior, say at $x \to +\infty$, of some linear transform

$$F(x) = \int K(x, y)f(y) dy$$

of a function $f(x)$ and wants to deduce the asymptotic behavior at infinity of f itself. Wiener's theorem has the following Tauberian aspect: if $f \in L^1$, its Fourier transform never vanishes and

$$\lim_{x \to +\infty} \int f(x - y)g(y) dy = A \int f(x) dx$$

for some $g \in L^\infty$, then the same holds for all $f \in L^1$. In fact, the limit relation above holds by an integration if f is replaced by any convolution

$$f * h(x) = \int f(x - z)h(z)dz,$$

where h is continuous with compact support and then, by taking the limit when f is replaced by any function in L^1. Then, if g is suitably regular at ∞, we may conclude that $g(t)$ tends to A for large t.

The point of Wiener's paper (1932) is to deduce a great number of Tauberian theorems from Wiener's own theorem. One of them implies the prime number theorem, which, however, is proved more simply by Ikehara's theorem: if $d\mu(x) \geq 0$ is a measure, if the integral

$$F(s) = \int_0^\infty e^{-sx}d\mu(x)$$

converges for $\operatorname{Re} s > 1$, if $F(1 + it) \neq 0$ when t is real and not zero and if

$$F(s) \sim A/(s - 1)$$

at $s = 1$, then $\mu(t)e^{-t}$ tends to A as $t \to \infty$.

Our interest will be Wiener's theorem in itself, properly named Wiener's density theorem. In this lecture, I shall reproduce what is essentially Wiener's original proof and compare it to the later, similar results in the theory of Banach algebras and in abstract harmonic analysis.

Generalities about the Fourier transform

The Fourier transform $\mathcal{F}$ on the real line has the inverse $\mathcal{F}^{-1}$ defined by

$$\mathcal{F}^{-1}g(x) = \frac{1}{2\pi}\int e^{ixt}g(t)dt,$$

for suitable classes of functions. If, for instance, $f \in L^2$ is smooth and small at infinity, it is well known that the same holds for $F = \mathcal{F}f$ and we have $f = \mathcal{F}^{-1}F$. For this class of functions the following inequalities are proved by integrations by parts:

$$|f(t)| \leq \int |F(x)|dx, \quad |tf(t)| \leq \int |dF(x)|, \quad f = \mathcal{F}^{-1}F;$$

in particular,

$$(1 + t^2)|f(t)| \leq \int (|F(x)|dx + |dF'(x)|).$$

By a passage to the limit this subsists for certain integrable functions f and all functions F in the class B_0 of continuous functions with compact supports and second derivatives of bounded variation. Hence B_0 is contained in TL^1.

It is clear that the class B_0 contains all piecewise linear functions with compact supports, in particular also the piecewise linear function $P(x)$ which equals 1 when $|x| \leq 1$ and vanishes when $|x| \geq 2$. Note that the sum

$$\sum_{-N}^{N} P(x + 2k)$$

equals 1 when $|x| \leq 2N$. The function $P(x)$ and its translates will play a key role in the sequel. Since each of them equals 1 on some interval, they will be called *local units*.

The perfect situation occurs for $L^2(R)$. By Plancherel's theorem $\mathcal{F}$ is unitary map on this space. In the theory of distributions it is proved that $\mathcal{F}$ is a bijection on the space of tempered distributions.

For the space $L^1 = L^1(R)$ that Wiener considered, the situation is less symmetrical. All that can be said of the space TL^1 is that its elements are uniformly continuous and vanish at infinity.

The space L^1 is a Banach space with the norm

$$\|f\| = \int |f(x)| dx.$$

Its dual is the space L^∞ of essentially bounded functions. The crucial property of L^1 is that it is an algebra under convolutions

$$f * g(x) = \int f(x - y) g(y) dy, \quad \|f * g\| \leq \|f\| \, \|g\|,$$

and we have

$$\mathcal{F}(f * g) = \mathcal{F}f \mathcal{F}g, \quad \mathcal{F}(e^{ixa} f(x)) = \mathcal{F}f(x - a).$$

All this follows from Fubini's theorem. Hence the transform $A = \mathcal{F}L^1$ is a ring under pointwise multiplication. As before, its elements will be denoted by capital letters, $F = \mathcal{F}f$ and so on. If we transport the norm on L^1 to $A = \mathcal{F}L^1$ so that $\|F\| = \|f\|$, we get

$$\|FG\| = \|F\| \, \|G\|.$$

Multiplication by an exponential, $f(t) \to e^{iat} f(t)$, is a linear isometry of L^1 and translation $(T_a F)(x) = F(x + a)$ is a linear isometry of A.

The subset A_0 of A whose elements have compact supports, in particular the space B_0 above, will play an important part in the sequel.

LEMMA 1. A_0 *is dense in* A.

PROOF. Let $0 \neq F \in A_0$. Then all the translates of F are in A_0 and also all products $e^{ita} F(t)$ for real a. Hence, if $g \in L^\infty$ is orthogonal to $T^{-1} A_0$, we have

$$\int e^{itb} f(t - a) g(t) dt = 0$$

for all real b and a. It follows that $f(t - a) g(t) = 0$ for almost all f and a countable dense set of values of a. Since $f \neq 0$, this shows that $g(t) = 0$ almost everywhere. Hence A_0 is dense in A.

Proof of Wiener's theorem

In view of the preceding lemma, Wiener's theorem now takes the following form in which it will be proved:

THEOREM W. *If* $F \in A$ *and* $F(x) \neq 0$ *for all* x, *then* AF *is dense in* A.

Note that if F has a zero x_0, then all elements of AF vanish at x_0 and hence the closure of AF is not equal to A.

The proof depends on the following

LEMMA. *Suppose that $F \in A$ and that $G \in B_0$. Then*

$$\|(F(x) - F(0))G(x/\varepsilon)\|$$

tends to zero as $\varepsilon > 0$ tends to zero.

PROOF. The function $G(x/\varepsilon)$ is the Fourier transform of $\varepsilon g(t\varepsilon)$. Hence $(F(x) - F(0))G(x/\varepsilon)$ is the Fourier transform of

$$\int f(s)\varepsilon g(\varepsilon(t-s))ds - \varepsilon g(\varepsilon t) \int f(s)ds.$$

After a change of variables $t \to t\varepsilon$, its norm is

$$\int dt \int |f(s)(g(t - \varepsilon s) - g(t))|ds.$$

By dominated convergence the double integral tends to zero as $\varepsilon \to 0$.

PROOF OF THE THEOREM. Let $P(x)$ be the local unit defined above, i.e. $P(x)$ is a piecewise linear function which vanishes when $|x| > 2$ and equals 1 when $|x| < 1$. Consider the quotient

$$\frac{P(x/\varepsilon)}{F(x)}.$$

Note that, by assumption, $F(x) \neq 0$ for all x. When $P(x/\varepsilon) \neq 0$, then $P(x/2\varepsilon)$ equals one and hence we may rewrite the quotient as

$$\frac{P(x/\varepsilon)}{F(0) + P(x/2\varepsilon)(F(x) - F(0))}.$$

By the lemma, the norm of

$$G_\varepsilon(x) = P(x/2\varepsilon)(F(x) - F(0))$$

tends to zero with ε. Hence, if ε is small enough, we can express our last quotient as a geometric series. The result is that

$$P(x/\varepsilon) = (F(x)/F(0)) \sum_{0}^{\infty} (-1)^k (G_\varepsilon(x)/F(0))^k$$

where the series converges so that the right side belongs to A. In other words, $P(x/\varepsilon) \in F(x)A$ when $\varepsilon > 0$ is sufficiently small. Since $|F(x)|$ has a positive lower bound on every compact interval, all translates $P(\frac{x+b}{\varepsilon})$ of $P(x/\varepsilon)$ belong to FA when b is bounded and ε is sufficiently small.

Now $P(x/\varepsilon)$ equals 1 in the interval $|x| \leq \varepsilon$ and a sum of suitable translates will be 1 in any interval. Hence to every interval $|x| \leq N$ there is a function in FA which equals 1 on this interval. But this obviously means that any element of A_0 belongs to FA. This finishes the proof of Wiener's Tauberian theorem.

Wiener's theorem for Fourier series

Let $L^1(Z)$ be the space of sequences $f = (f(n))$ labelled by the integers n and such that

$$\|f\| = \sum |f(n)| < \infty.$$

The Fourier transform $F = \mathcal{F}f$ of f is then a continuous function

$$F(x) = \sum f(n)e^{-inx}$$

in the interval $|x| < \pi$ such that $F(-\pi) = F(\pi)$. We can also consider $F(x)$ to be a continuous function on the unit circle. Hence the previous real line is replaced by a compact set. Let A be the space of these functions with the norm

$$\|F\| = \|f\|.$$

In other words, A is the space of 2π-periodic functions with convergent Fourier series. In contrast to the previous case, the ring A now contains a unit 1.

We can now repeat the previous argument with the functions $P(x/\varepsilon)$ and prove Wiener's result that $1/F(x)$ has a convergent Fourier series when $F(x) \neq 0$ everywhere.

Remark. Actually, Wiener used a different argument based on the fact that if

$$|f(0)| > \sum_{n \neq 0} |f(n)|,$$

then $F(x) \neq 0$ everywhere and $1/F(x)$ can be written explicitly as a function in A. He then showed that if $F(0) \neq 0$, the constant term dominates in the same way the Fourier series of

$$F_\varepsilon(x) = P(x/\varepsilon) + F(0)(1 - P(x/\varepsilon))$$

when $\varepsilon > 0$ is sufficiently small. Hence $1/F_\varepsilon(x)$ belongs to A. Since $F_\varepsilon(x) = 1$ close to 0, Wiener could then complete the proof by the same arguments as above.

Remark. Wiener's proof of the density theorem for Fourier integrals uses the results for Fourier series, now for functions which are periodic in a large interval I and vanish close to the endpoints. In this case the Fourier series for such functions approximate their Fourier integrals when the interval tends to the entire real line.

Normed rings

The space $L^1(Z)$ is a Banach space with norm $\|f\|$ and also a commutative ring under convolution $f * g$ such that $\|f * g\| \leq \|f\| \, \|g\|$. Moreover, it has a unit e given by $e(n) = \delta(n)$. We have a Banach space which is also a commutative ring with a unit.

In his influential articles (1941), I. Gelfand investigated such objects which he called normed rings, i.e. rings $R = (e, a, b, c, \dots)$, which are complex Banach spaces with a norm $|a|$, a commutative multiplication $ab = ba$ such that $|ab| \leq |a||b|$, and a unit e of norm one. The geometric series

$$\frac{e}{a - b} = a^{-1} \sum_0^\infty b^k a^{-k},$$

where a is invertible and b is small, shows that the set E of invertible elements is open. If R is a field, i.e. all elements except zero are invertible and an element a is not a complex multiple of e, then

$$(e + za)^{-1}$$

is an entire analytic function with values in R and hence a constant which must be zero. This contradiction shows that a normed field consists of all complex multiples of the unit.

In the general case, the main trick is now to consider ideals of R, by definition not equal to R. Any such ideal does not intersect the ball $E : |e - a| < 1$ and hence is contained in the closed set $R - E$. It follows that the closure of an ideal is an ideal and so is the union of any ascending chain of ideals. Hence we have the following observation which relies heavily on the existence of a unit, namely

Every ideal is contained in a maximal ideal.

Here an ideal I is said to be maximal when there is no ideal strictly between I and R. If I is a maximal ideal, the quotient $x = R/I$ is a field and conversely. Hence the quotient map $a \to a(x)$ can be identified with a homomorphism $a \to a(x)$ into the complex numbers such that

$$(ab)(x) = a(x)b(x), \quad |a(x)| \le |a|,$$

where the ideal I consists of all a such that $a(x) = 0$.

We can now define a maximal ideal space X as a set of points x, each one corresponding to a maximal ideal. The first main theorem of Gelfand's theory is now

THEOREM. *An element $a \in R$ is invertible if and only if $a(x) \ne 0$ for all x in the maximal ideal space.*

In fact, Ra is then not contained in any maximal ideal and must be R itself.

Application to absolutely convergent Fourier series

When $R = L^1(Z)$, what is the maximal ideal space X? To answer this question, let us note that $L^1(Z)$ has a generator g of norm 1, with an inverse of norm 1 given by $g(n) = \delta(n - 1)$ with the inverse $g^{-1}(n) = \delta(n + 1)$. The elements of R have the form

$$a = \sum c_n g^n, \quad |a| = \sum |c_n|.$$

If $x = R/I$ and I is a maximal ideal, it follows that $|g(x)| \le 1$ and $|g^{-1}| \le 1$. Hence $|g(x)| = 1$ and

$$a(x) = \sum c_n g(x)^n$$

is indeed a homomorphism of R into the complex numbers. Writing $a(x) = e^{it}$ we have recovered the absolutely convergent Fourier series and Wiener's theorem.

The convolution ring of absolutely convergent measures also has a unit, and here Gelfand's methods give a very simple proof of a theorem by Wiener and Pitt, which says that a measure is invertible if the total measure of its singular part is less than the lower bound of the absolute values of its Fourier transform. For $L^1(R)$, which is a ring without a unit, this simple reasoning with maximal ideals does not work and additional arguments are necessary. We shall not go into this nor shall we comment on the rest of Gelfand's papers listed in the bibliography. Instead we shall return to $L^1(R)$ and Wiener's theorem but now with a new angle and with methods which apply to R^n and other groups.

Fourier analysis on locally compact Abelian groups

The possibility of extending Fourier analysis from the real line to locally compact abelian groups was pointed out by André Weil in a very influential book (1938). If $G = (x, y, \dots)$ with addition $x + y$ and inverse $-x$ is such a group, it has a measure $\mu \geq 0$ which is finite on compact subsets C and invariant in the sense that $\mu(C + y) = \mu(C)$. We can then form the space $L^1(G)$ of integrable functions with finite norm

$$\|f\| = \int_G |f(x)| d\mu(x).$$

For such functions there is a commutative and associative convolution $f * g$. The dual group Γ is the space of characters, continuous homomorphisms $x \to (x, \xi)$ of G into the unit circle U. Examples: if $G = R^\mu$, $\Gamma = R^\nu$, if $G = Z^\nu$, $\Gamma = U^\nu$, if G is discrete, Γ is compact.

The Fourier transform

$$\mathcal{F}f = \int f(x) e^{i(x, \xi)} d\mu(x)$$

then maps $L^1(G)$ into the subspace of $L^\infty(\Gamma)$ whose elements vanish at infinity in Γ. The ordinary Fourier integral and Fourier series are special cases of this theory. In the forties, many efforts were made to extend classical harmonic analysis to the general framework of locally compact abelian groups.

Extension of Wiener's theory

We shall extend Wiener's theory to locally compact abelian groups. To simplify, we reason with $G = R^\nu$ and $\Gamma = \hat{R}^\nu$ in such a way that the results extend to the general case.

To begin with, we shall construct local units.

Local units and their decomposition

In the sequel we shall write the Fourier transform of $f \in L^1 = L^1(G)$ as

$$\hat{f}(\xi) = \int e^{ix\xi} f(x) dx, \quad x\xi = \sum x_k \xi_k,$$

and differentiate between $G = (x, y, \dots)$ and $\Gamma = (\xi, \eta, \dots)$, connected by the duality $x\xi$. Let A be the set of Fourier transforms of functions in L^1.

A local unit is now a function in L^1 whose Fourier transform has compact support and equals 1 on some non-empty open set $U \subset \hat{R}^n$.

LEMMA 1. *Given a bounded open non-empty subset $V \in \hat{R}^\nu$, there is a local unit $F \in L^1$ such that $\hat{F}$ is supported in V and equals 1 on a neighborhood U of the origin. Moreover,*

(1) $$\|F\| \leq 2c, \quad c = (2\pi)^n,$$

and

(2) $$\|(T_x - 1)F\| \leq 4c \sup_V |(e^{ix\xi} - 1)|$$

where T_x is translation by x, $(T_x F)(y) = F(x + y)$.

Note that (2) means that $F(x)$ is close to 1 on any compact set when V is small.

PROOF. There are open sets W such that $\bar{W} \subset V$ and hence also an open neighborhood U of the origin such that $W \pm U \subset V$. Let $|V|$ be the volume of V and similarly for $|W|$ and choose W such that $4|W| > |V|$. Let $\hat{f}, \hat{g}$ be the characteristic functions of V and W. We shall see that

$$F(x) = f(x)g(x)/|W|$$

has the required properties. In fact $\|f\|_2 = \sqrt{V}$ and $\|g\|_2 = \sqrt{W}$ so that by Parseval's formula,

$$\|F\| \le \|f\|_2 \|g\|_2/|W| \le c\sqrt{|V|/|W|} \le 2c$$

and, since

$$\hat{F}(\xi) = \int \hat{f}(\xi - \eta)\hat{g}(\eta)d\eta,$$

$\hat{F}$ is supported in V and equals 1 on U. Also

$$|W| \|(T_x - 1)F\| \le \|(T_x - 1)f\|_2 \|g\|_2 + \|f\|_2 \|(T_x - 1)g\|_2.$$

Since the Fourier transform of $(T_x - 1)f$ is $(e^{ix\xi} - 1)\hat{f}(\xi)$, insertions of $\|f\|_2$ and $\|g\|_2$ prove (2).

The proof of the next lemma is obvious.

LEMMA 2. *If U is an open set covered by a union of finitely many open sets $U_1, \ldots, U_m$, and $F_1, \ldots, F_m$ are local units such that F_k equals 1 on U_k, then*

$$1 - (1 - \hat{F}_1) \cdots (1 - \hat{F}_m)$$

is a local unit equal to 1 on U.

Modules for L^1 and Beurling's theorem

The dual of $L^1 = L^1(G)$ is L^∞ and if $g \in L^\infty$, $f \in L^1$, the convolution

$$f * g \in L^\infty$$

is well defined. In other words, L^∞ is an L^1-module.

Let us now define *spectra* and *null sets* for functions in L^1 and L^∞.

DEFINITION. *A point $\xi \in \Gamma$ belongs to the spectrum $\mathrm{sp}(g)$ of a function $g \in L^1(G)$ or $L^\infty(G)$ if, given any neighborhood N of ξ, there is $f \in L^1$ such that $\hat{f}$ is supported in N and $f * g \ne 0$. If, on the contrary, $f * g = 0$ under the same conditions, ξ belongs to the null set $N(g)$ of g.*

Remark. It follows from the definition that $\mathrm{sp}(g)$ and $N(g)$ are closed but that their interiors have an empty intersection. When $g \in L^1$, the Fourier transform of $f * g$ is $\hat{f}\hat{g}$ and hence $\mathrm{sp}(g)$ is simply the support of $\hat{g}$. When $g \in L^\infty$, $\mathrm{sp}(g)$ is also the support of the Fourier transform of g, regarded as a distribution. But since our arguments are meant to apply to locally compact abelian groups, we cannot use the notion of distribution. Still we may imagine that $\mathrm{sp}(g)$ is the support of its virtual Fourier transform.

Multiplication by an exponential translates a spectrum. In fact, the Fourier transform of $e^{ix\eta}g(x)$ is $\hat{g}(\xi-\eta)$ when $g \in L^1$ and a small computation gives the same result when $g \in L^\infty$. Similarly: a translation in G corresponds to multiplication by an exponential in Γ.

Beurling's theorem

We can now prove a slight extension of a theorem by Beurling (1945) which implies Wiener's theorem.

THEOREM B. *Suppose that $g \in L^\infty$ and that $\xi \in \mathrm{sp}(g)$. Suppose further that F_n is a bounded sequence of local units whose Fourier transforms equal 1 close to ξ and such that their spectra tend to ξ. Then all the limits of the functions $F_n * g$ under locally uniform convergence are multiples of $e^{ix\xi}$.*

Remark. This implies Theorem W, for if the Fourier transform of $f \in L^1$ never vanishes and $f * g = 0$ for some $0 \neq g \in L^\infty$, then $\mathrm{sp}(g)$ must contain a point ξ so that $\hat{f}(\xi) = 0$ which is a contradiction. Hence $f * g = 0$ implies that $g = 0$ so that $L^1 * f$ must be dense in L^1. But Beurling's theorem is a bit stronger.

Remark. In the original paper, Beurling distanced himself from Wiener by phrasing his theorem as a property of bounded and uniformly continuous functions $f(x)$ on the real line, namely the following: if f does not vanish identically, then there are an exponential e^{ixt} and linear combinations of translates,

$$g(x) = \sum c_k f(x - x_k),$$

which tend to the exponential uniformly on compact sets and so that $\sup |g(x)|$ tends to 1. In the proof below and under the only condition that $f(x)$ is essentially bounded and not zero, there are convolutions

$$\int f(x - y)g(y)dy, \quad g \in L^1$$

which tend to the exponential in the manner indicated. When $f(x)$ is also uniformly continuous, such integrals can be approximated in the same way by linear combinations above.

PROOF. By a translation of $\mathrm{sp}(f)$, we may assume that $\xi = 0$. Let G_n be local units constructed in Lemma 1 such that their spectra tend to 0 and $G_n * F_n = F_n$. Then no $g_n = G_n * g$ vanishes, the sequence $\|g_n\|$ is bounded, $\mathrm{sp}(g_n)$ tends to zero and, by the lemma,

$$|g_n(x) - g_n(0)| = |F_n * g_n(x) - F_n g_n(0)| \leq \|(T_x - 1)F_n\| \, \|g\|_\infty$$

tends to zero locally uniformly as $n \to \infty$. Hence every subsequence of (g_n) has a subsequence $(g_{n'})$ for which $g_n(0)$ converges and hence $g_{n'}(x)$ is a bounded sequence of functions which tends to a constant c, uniformly on compact subsets.

Remark. If we normalize so that $\|g_n\|_\infty = 1$ and translate g_n so that, for instance,

$$1 \geq g_n(0) \geq 1/n,$$

this sequence has norm 1 and tends to 1 locally uniformly. Note that if $\mathrm{sp}(g)$ consists of 0 alone, all $g_n(x)$ are the same and hence equal to some constant $\neq 0$.

Spectra and null sets of ideals. Two ideals with the same null set

The null set $N(I)$ of an ideal I is defined as the intersection of all $N(f)$ for $f \in I$; the spectrum $\text{sp}(I)$ is the union of all $\text{sp}(f)$. Both these subsets of Γ are closed and the same for an ideal and its closure in $L^1(G)$.

After these definitions, the following problem seems natural: are there two different closed ideals with the same null set? In 1948 Laurent Schwartz proved very simply that the answer is no when the null set is the unit sphere E in $\Gamma = R^3$. In fact, the inverse Fourier transform of the unit rotation invariant measure $d\mu(\xi)$ on E turns out to be $\sin|x|/|x|$. Hence, if f and $\hat{f}$ are smooth functions, then

$$\int f(x) \frac{x_1}{|x|} \sin|x| dx = \text{const} \int \frac{\partial \hat{f}(\xi)}{\partial \xi_1} d\mu(\xi).$$

The left side is a continuous function of $f \in L^1$ which vanishes on the closure of the set I of all smooth functions f such that

$$\hat{f}(\xi) = 0, \quad \frac{\partial f(\xi)}{\partial \xi_1} = 0$$

on E but does not vanish on the set J of smooth functions which satisfy just the first equality. Moreover, both I and J are translation invariant. Hence $N(\bar{I}) = N(\bar{J}) = E$ but $\bar{I}$ and $\bar{J}$ are not the same ideal.

When does an element belong to an ideal?

After the counterexample above, it seems interesting to know general criteria involving spectra which guarantee that a given function belongs to a given ideal. We shall prove one such theorem which follows from Beurling's theorem.

THEOREM. *If $I \subset L^1$ is a closed ideal, $f \in L^1$ and $N(I) \cap \text{sp}(f)$ is countable, then $f \in I$.*

Remark. Since $N(I)$ and $\text{sp}(f)$ are closed, their intersection is a closed, countable set.

PROOF. By duality it suffices to prove that

$$I * g = 0 \implies f * g = 0$$

for every $0 \neq g \in L^\infty$. By the first equality, the spectrum of g has an empty intersection with the interior of the spectrum of I. In other words, $\text{sp}\,g \subset N(I)$. By assumption, $N(I) \cap \text{sp}(f)$ must have isolated points ξ. Hence if H is a local unit with support close enough to ξ, the function $H * f * g$ is actually independent of H, i.e. $H_1 * f * g = H_2 * f * g$ when H_1, H_2 are as H. By Theorem B, there is a sequence H_n of local units whose spectra tend to ξ such that the limits of $H_n * g$ are multiples of $e^{ix\xi}$ and hence the limit of $H_n * f * g$ must be a multiple of $\hat{f}(\xi) = 0$. Hence $H * f * g = 0$ when H is a local unit whose spectrum is close enough to ξ.

Since every subset of $N(I) \cap \text{sp}(g)$ must have isolated points, it follows that $H * f * g$ vanihes for all local units with spectra close to $N(I) \cap \text{sp}(g)$. Hence the same holds for all local units and hence $f * g = 0$.

Permitted sets

When $C \subset \Gamma$ is closed, let $I(C)$ be the ideal of functions $f \in L^1(G)$ such that $\hat{f}$ vanishes on C. Let us say that C is permitted if $I(C)$ is the only ideal whose null set is C.

The theorem just proved shows that a finite collection of points in G or a finite collection of closed intervals in R is permitted but not much more. But many more sets are permitted, at least in $\Gamma = \hat{R}^n$, for instance radial sets with the property that they contain a point ξ such that

$$\bigcap_{a \geq 1} a(C - \xi) \supset C - \xi$$

for some ξ. In the proof we can take $\xi = 0$. Suppose that $\hat{f}$ vanishes on C. Then $\hat{f}(\xi/a)$ vanishes in a neighborhood of C and is the Fourier transform of $f_a(x) = a^n f(ax)$ which belongs to $I(C)$. Since $\|f_a - f\|$ tends to zero as a decreases to 1, f itself belongs to $I(C)$.

We have now come to the end of our journey from Wiener's theorem to abstract harmonic analysis, once a subject of intense interest. Rudin's book (1962) contains among many other things a proof of Malliavin's negative result: there are closed sets in every non-discrete abelian group which are not permitted.

Bibliography

BEURLING A.

    1945. *Un théorème sur les fonctions bornées et uniformément continues sur l'axe réel*, Acta Math. **77** (1945), 127–136.

GELFAND I.

    1941. a. *Normierte Ringe*, Mat. Sb. **9(51)** (1941), 1–24.

    1941. b. *Über verschiedene Methoden der Einführung der Topologie in die Menge der maximalen Ideale*, (with Schilow), l.c., 25–40.

    1941. c. *Ideale und primäre Ideale in normierten Ringen*, l.c., 41–48.

    1941. d. *Zur Theorie der Charaktäre der Abelschen topologischen Gruppen*, l.c., 49–50.

    1941. e. *Über absolut konvergente trigonometrische Reihen und Integrale*, l.c., 51–66.

MALLIAVIN P.

    1959. *Sur l'impossibilité de la synthèse spectrale sur les groupes abéliens non compacts*, Publ. Math. IHES (1959), 61–68.

RUDIN W.

    1962. *Fourier analysis on groups*, Interscience tracts in Pure and Applied Math. no. 12, 1962.

WEIL A.

    1938. *L'intégration dans les groupes topologiques et ses applications*, Gauthiers-Villars, Paris, 1938.

WIENER N.

    1932. *Tauberian theorems*, Annals of Mathematics **33** (1932), 1–94.

CHAPTER 7

The Tarski-Seidenberg Theorem

Introduction

A semialgebraic set is a real subset of some R^n defined by real polynomial inequalities. The Tarski-Seidenberg theorem, which says that projections of semialgebraic sets are also semialgebraic, is of general interest and has found important applications in the theory of general partial differential operators, in particular the hyperbolic and hypoelliptic ones to be treated in the next two lectures. The purpose of this one is to give a leisurely proof of the Tarski-Seidenberg theorem. The original proof by the logician Tarski was edited by Seidenberg (1954) in a more informal way. Another unpublished logical version due to Paul Cohen (1967) was rephrased by Lars Hörmander (1983, II, Appendix 2, p. 364). His version is followed here.

Some examples

The lecture starts with examples of some elementary steps in elimination theory which are used over and over again in the proof of the Tarski-Seidenberg theorem.

Do two real, given polynomials have a common real zero?

Consider real non-constant polynomials $f(x), g(x), \ldots$ in one real variable. Let $Q(f, g)$ be the following question: do f and g have a common real zero? To simplify the question, suppose that $\deg f \geq \deg g$ and write $f = hg + k$, $\deg k < \deg g$. Then it is obvious that $Q(f, g)$ and $Q(g, k)$ have the same answers. Continuing in this way, we are reduced to a question $Q(F, G)$ where $\deg F > 0$ and G is a constant. In this case the answer is no if $G \neq 0$, yes if $G = 0$ and $\deg F$ is odd. If the degree of F is even and positive, the answer is in doubt. We shall see below that the way out in this case is to consider the pair $Q(F, F')$.

Do three, four, ... real, given polynomials have a common zero?

Take for instance three polynomials f_1, f_2, f_3 and let $Q(f_1, f_2, f_3)$ be the question. It remains the same if the pair (f_2, f_3) is reduced as above to a pair (F_2, G_3) where G_3 is a constant. If $G_3 \neq 0$, the answer is no, otherwise we are reduced to the question $Q(f_1, F_2)$ which can be reduced as above to the case of just one polynomial of even degree. An answer in this case can be found by a procedure outlined below.

The sign

In the examples above, the actual positions of eventual zeros are irrelevant. To decide similar questions in more general cases, it is convenient to formalize the sign properties of a real polynomial $f(x)$ by the symbol $\mathrm{SGN}\, f$ which notes the successive zeros of the polynomial and its signs in the intervals between. In the text we shall use the word sign in this sense. A constant can have one of three signs $+, 0, -$ and a first degree polynomial can only have the signs $-0+$ or $+0-$. The only possible signs of a second degree polynomial are

$$+, \quad -, \quad +0+, \quad -0-, \quad +0-0+, \quad -0+0-$$

depending on whether it has no, one or two zeros. The number of possibilities increases rapidly with the degree.

Following Lars Hörmander, we shall also introduce the combined sign of m polynomials

$$\text{SGN}(p_1, \ldots, p_m)$$

which registers, in increasing order, all the zeros of these polynomials and the signs of all polynomials at each zero and every interval between, including the intervals stretching to $\pm\infty$.

Example. Let $p(x) = x^2 - 1, q(x) = x + 3$. Then

$$\text{SGN}(p, q) = +/-, \quad +/0, \quad +/+, \quad 0/+, \quad -/+, \quad 0/+, \quad +/+$$

where the sign above / refers to p and the sign below / refers to q. It is clear that the complexity increases a lot with the number of polynomials.

LEMMA 1. *When $f(x)$ is a real polynomial and $r(x)$ is the remainder when $f(x)$ is divided by $f'(x)$, then $\text{SGN}(f)$ is uniquely determined by $\text{SGN}(f', r)$.*

Example. Let $f(x) = ax^2 + 2bx + c$ with $a > 0$. Then $f'(x) = 2ax + 2b$ and the remainder is $r(x) = c - b^2/a$. The combined sign of (f', r) is

$$-/e, \quad 0/e, \quad +/e$$

where $e = \text{sgn}\, r$. If $e = 0$, $f(x) \geq 0$ with just one zero and $\text{SGN}\, f = +0+$. If $e = +$, $f(x)$ is positive definite with the sign $+$ and if $e = -$, the minimal value of f is negative and f has the sign $+, 0, -, 0, +$.

PROOF. Let $f(x) = ax^n + \cdots$ have degree n. Changing f to $-f$ changes all signs of f and (f', r) and hence we may assume that $a > 0$. This choice is also apparent in $\text{SGN}\, f'$ and determines the signs of f in the end intervals at $\pm\infty$. The signs of f' at the corresponding intervals determine whether f increases or decreases in these intervals. At the first zero x_0 of $f'(x)$, the sign of $r(x_0)$ determines the sign of $f(x_0)$. Hence $\text{SGN}(f)$ is determined when $x \leq x_0$. At the next zero x_1 of $f'(x)$, the sign of the remainder determines the sign of $f(x_1)$ and hence also, since $\text{sgn}\, f(x_0)$ is known and $f(x)$ is monotone between x_0 and x_1, the sign of $f(x)$ between x_0 and x_1 which may or may not include a zero. The induction is now obvious and the lemma is proved. Note that only the signs of the remainder at the zeros of $f'(x)$ are needed to reconstruct the sign of $f(x)$. Hence $\text{SGN}(f)$ only determines $\text{SGN}(f', r)$ within a finite number of possibilities.

We shall also prove a similar lemma.

LEMMA 2. *Let p, q be real polynomials in one variable such that $\deg q > 0$ and $\deg q \geq \deg p$. Then $\mathrm{SGN}(p, q)$ is determined by $\mathrm{SGN}(p, q', r, s)$ where r is the remainder when q is divided by q' and s the remainder when q is divided by p.*

Remark. The proof shows that one has the same result when p is replaced by $p_1, \ldots, p_j$ and $s_1, \ldots, s_j$ are the corresponding remainders.

PROOF. By the previous lemma, $\mathrm{SGN}(q', r)$ determines $\mathrm{SGN}(p)$. The relative position of $\mathrm{SGN}(p)$ to $\mathrm{SGN}(q)$ is then determined by the values of s at the zeros of q in the monotonicity intervals of q. Since $\mathrm{SGN}(q', s)$ is known, the positions and signs of these values are known.

Semialgebraic sets

A *semialgebraic set* is a set of points $x \in R^n$ which satisfy some subset of a set of inequalities

$$p(x) = 0, \ q(x) \geq 0$$

where p and q each stand for a finite set of polynomials. The opposite inequalities are taken care of by changing the signs of the polynomials. Note that the definition means that a semialgebraic set has a finite number of components consisting of sets of points x where all inequalities of a subset of the ones above are satisfied simultaneously. It follows that finite unions and intersections of semialgebraic sets are again semialgebraic. A semialgebraic set on the line is a finite collection of intervals which may be points.

Example. Let $f(x) = x^2 + 2ax + b$ and $g(x) = cx + d$, $c > 0$. We shall see that $f(x) = 0$, $g(x) > 0$ holds for a semialgebraic set in the coefficients a, b, c, d. In fact, if $f(x) > 0$ for all x, i.e. if

$$b > a^2,$$

there is no condition on c, d except the original $c > 0$. If $f(x) \geq 0$ has a zero, we get

$$b = a^2, \quad g(-a/2) > 0.$$

If $f(x) = 0$ has two zeros, the zero $-d/c$ of $g(x)$ is either to the left or the right of the interval where $f(x) < 0$ and we have

$$b < a^2, \quad f'(-d/c) < 0 \quad \text{or} \quad f'(-d/c) > 0.$$

In the last two inequalities we can multiply by c and hence our three formulas together define the desired semialgebraic set.

The Tarski-Seidenberg theorem

The Tarski-Seidenberg theorem says that projections of semialgebraic sets are semialgebraic. The proof uses the notion of SGN and the tricks already exhibited in the examples.

TARSKI-SEIDENBERG. *The projection of a semialgebraic set is semialgebraic.*

Remark. This means for instance that if $P_1(x, y), \ldots, P_t(x, y)$ are real polynomials in $x \in R^m, y \in R^n$, then all y for which $P_1, \ldots, P_t$ have fixed signs form a semialgebraic set.

PROOF. We shall proceed by an induction according to a very complicated branching system. To begin with, it suffices to take $m = 1$ and just one x variable. Then there are only a finite number of possibilities for $\mathrm{SGN}(P_1, \ldots, P_t)$ with a fixed y. Hence it suffices to consider semialgebraic sets

$$E : \mathrm{SGN}(P_1, \ldots, P_t) = w,$$

with given y and w. We shall proceed by the rank of $P_1, \ldots, P_t$ meaning the highest degree and the number of polynomials with that degree. One more induction will appear in the course of the proof. Suppose that

$$P_1 = a_1(y)x^m + \cdots, \qquad P_j = a_j(y)x^m + \cdots, \qquad m > 0,$$

have the highest degree in x and let $a_1(y), \ldots, a_r(y)$ be all the highest coefficients appearing in the polynomials $P_1, \ldots, P_t$ of positive degree. Add to the rank also a sum, more precisely the sum of the degrees of the polynomials not of maximal degree. Our semialgebraic set is then the union of the set where no $a_1(y), \ldots, a_r(y)$ vanishes and semialgebraic sets of lower rank or the same rank and lower sum. If the rank and sum vanish there is nothing to prove. Hence we may proceed by induction from the intersection I of E with the set where $a_1(y), \ldots, a_r(y)$ have fixed signs, not equal to zero.

By the remark of Lemma 2, I is implied by a semialgebraic set of lower rank and sum (when the highest coefficients appear in denominators, just multiply them away by their squares). This finishes the proof.

The theorem says that if $x \in R^n$, $y \in R^m$, then the set

$$(x; \exists y \, (x, y) \in S)$$

is semialgebraic when $S \subset R^{n+m}$ is semialgebraic. Hence, if S' is the complement of S, also

$$(x; \forall y \, (x, y) \in S')$$

is semialgebraic. By superposing $\exists$ and $\forall$ we may prove that the closure

$$(x; \forall \varepsilon > 0 \, \forall j \, |x_j - y_j| < \varepsilon \, \exists x \in S)$$

of a semialgebraic set $S \in R^n$ is semialgebraic and hence also the boundary of such a set.

Applications

Let $S \in R^{2+n}$ be a semialgebraic set with the variables x, y, z. Then the semialgebraic set

$$T : ((x, y); \forall \varepsilon > 0 \, \exists y' > y - \varepsilon, \, (x, y', z) \in S)$$

has the property that its intersections with $x = $ const are closed intervals. Hence it is the subgraph of a function $y = f(x)$ defined by $f(x) = \sup y$ for $(x, y) \in T$.

The function f may have the value $+\infty$ and we put it equal to $-\infty$ when it is not defined, a set easily seen to be semialgebraic.

LEMMA. *Let $f(x)$ be a real function of one variable with a semialgebraic subgraph. Then a finite number of points bound a finite set of intervals where f equals a locally analytic algebraic function.*

PROOF. Since its subgraph is semialgebraic, it is defined by a number of simultaneous inequalities picked from some set of polynomial inequalities and equalities

$$p_1(x, y) > 0, \ldots, q_1(x, y) = 0, \ldots, \quad (x, y) \in R^2.$$

The projections of this set on the x-axis and the y-axis are finite collections of intervals. Since the inequalities define an open set and the subgraph is not open, the set above includes equalities. The set of points x where the corresponding polynomials have only simple, separate zeros $y(x)$, is the complement of a finite set of points and consists of a finite set of intervals. Let I be one of them. In this interval, the equalities reduce to the zeros of a set of continuous algebraic functions $f_1(x), \ldots, f_m(x)$. By the properties of the subgraph, one of them must coincide with $f(x)$ in I. This finishes the proof.

If $f(x)$ exists $< \infty$ in one end interval, Puiseux's theorem shows that there

$$f(x) = A|x|^a(1 + O(1))$$

for some A and rational a. This fact will be used in the lectures on hyperbolicity and hypoellipticity.

Bibliography

HÖRMANDER L.
  1983. *The analysis of linear partial differential operators*, I, II, Grundlehren der Math. Wiss., vols. 256, 257, Springer, 1983.
SEIDENBERG A.
  1954. *A new decision method for elementary algebra*, Ann. Math. **60** (1954), 365–374.

CHAPTER 8

Intrinsic Hyperbolicity

Introduction

The classical hyperbolic differential operator is the wave operator

$$W(D) = D_1^2 - D_2^2 - \cdots - D_n^2, \quad D_k = \partial/i\partial x_k,$$

where x_1 is time. Its characteristic polynomial

$$W(\xi) = \xi_1^2 - \xi_2^2 - \cdots - \xi_n^2$$

has two real zeros in the first variable when the others are real. In the classical book (1931) by Courant and Hilbert homogeneous operators of order $m > 2$ were said to be 'totally hyperbolic' when their characteristic polynomials $P(\xi)$ have the same property: m real zeros in the first variable when the others are real. Below we shall get a better understanding of the situation.

One important object associated with the wave operator is the light cone

$$C : x_1 \geq 0, \ x_1^2 - x_2^2 - \cdots - x_n^2 \geq 0.$$

For the wave equation there is an explicit solution of Cauchy's problem: find a solution u of $W(D)u = 0$ with initial data $u = f$ and $D_1 u = g$ given for $x_1 = 0$. One essential and well-known property of the wave operator is that the value $u(y)$ of the solution only depends on the values of the initial data at a compact set, namely the intersection

$$(y \pm C) \cap (x_1 = 0).$$

In particular, if the initial data tend to zero uniformly on compact parts of the plane $x_1 = 0$, then the solution tends to zero locally uniformly.

Inspired by a paper by Petrovskii (1938) and by some early lectures on the theory of distributions, I worked out an intrinsic definition (1951) of hyperbolicity for partial differential operators $P(D)$ with constant coefficients. The definition requires that all smooth solutions u of the equation $P(D)u = 0$ shall tend to zero everywhere when all the derivatives of u up to some fixed order tend to zero uniformly in every compact part of some hyperplane $(x, N) = 0$. Let us call this class of differential operators $H(N)$. Using exponential solutions, this condition gives a necessary condition $\text{Hyp}(N)$ on the characteristic polynomial $P(\xi)$ of $P(D)$. This in turn permits the construction of a fundamental solution $E(P, x)$ of $P(D)$ which shows that $H(N) = \text{Hyp}(N)$.

In my paper I had to prove a special case of the Tarski-Seidenberg theorem, then nonexistent.

Intrinsic and algebraic hyperbolicity

As above we shall use the current notation for two dual spaces R^n

$$x = (x_1, \ldots, x_n) \in R^n, \quad \xi = (\xi_1, \ldots, \xi_n) \in R^n,$$

with a Latin-Greek duality $(x, \xi) = \sum x_k \xi_k$. Let $D = \partial/i\partial x$ be the imaginary gradient and write a partial differential operator with constant coefficients as $P(D)$ where

$$P(\xi) = \sum a_\alpha \xi^\alpha$$

is a polynomial. Suppose it has degree $m > 0$ which means that $|\alpha| \leq m$ above and that the principal part of P,

$$P_m(\xi) = \sum a_\alpha \xi^\alpha, \quad |\alpha| = m,$$

does not vanish. A vector $N \neq 0$ is said to be characteristic for P if $P_m(N) = 0$. Let me first repeat the definition of $H(N)$.

DEFINITION. *The operator $P(D)$ is said to be in the class $H(N)$, if all solutions u of $P(D)u = 0$ tend to zero locally uniformly in R^n when all their derivatives tend to zero locally uniformly in the plane $(x, N) = 0$.*

We shall also define a corresponding algebraic class.

DEFINITION. *A polynomial P is said to be in the class $\mathrm{Hyp}(N)$ if N is not characteristic for P and all the zeros $t = t(\xi)$ of the equation $P(tN + \xi) = 0$ have bounded imaginary parts for all real ξ.*

Remark. It is obvious that this class is closed under multiplication and differentiation in the N direction. A polynomial in $\mathrm{Hyp}(N)$ is said to be hyperbolic with respect to N. If we factorize

$$(1) \qquad\qquad P(tN + \xi) = P_m(N) \prod_1^m (t + \tau_k(\xi)),$$

the hyperbolicity condition says that $\sum |\operatorname{Im} \tau_k(t)| \leq \mathrm{const}$ for all real ξ. In particular, if $P = P_m$ is homogeneous, all zeros $\tau_k(\xi)$ are real for all real ξ. This is the classical condition and it is easy to prove (see Lemma 2 below) that $P \in \mathrm{Hyp}(N) \implies P_m \in \mathrm{Hyp}(N)$. A novelty at the time was that the converse is not true (see the end of the lecture).

Our aim is a proof of

THEOREM. *$P(D) \in H(N)$ if and only if $P \in \mathrm{Hyp}(N)$.*

Our first step is

LEMMA 1. *If $P(D) \in H(N)$, then $P \in \mathrm{Hyp}(N)$.*

PROOF. If $P = P_m$ is homogeneous, and N is characteristic, then $u(x) = f((x, N))$ solves the equation $P(D)u = 0$ for any $f \in C^\infty$. In fact,

$$P_m(D)u(x) = P_m(N/i)f^{(m)}((x, N)).$$

Hence, taking f not identically zero but $f(t) = 0$ close to $t = 0$ shows that P_m cannot belong to $\mathrm{Hyp}(N)$. In the general case, let ξ be non-characteristic and consider exponential solutions

$$(2) \qquad\qquad u = e^{i(x, tN + s\xi)}, \quad P(tN + s\xi) = 0,$$

with the values

(3) $$u_1 = e^{i(N,tN+s\xi)}, \quad u_2 = e^{is(x,\xi)}$$

when $x = N$ and $(x, N) = 0$, respectively. If $P(tN + s\xi)$ is independent of t, we may choose it positive and arbitrarily large in (2) which shows that P cannot be in $H(N)$. Assume next that N is characteristic. Then by the theory of Puiseux series, the equation $P(tN + s\xi) = 0$ has a non-zero solution $s = at^{1/p}(1 + o(1))$ for large $|t|$ and some integer $p > 1$. Then, with t as above and $it \to \infty$, $|u_1|$ tends to infinity much faster than the maximum of $(1 + |t| + |\xi|)^{c_1}|u_2|$ for $|x| \le c_2$ and this for any c_1, c_2. Hence P cannot be in $H(N)$ when N is characteristic.

Next assume that N is non-characteristic and consider solutions

$$u = e^{i(x,tN+\xi)}, \quad P(tN + \xi) = 0.$$

The maximum of the derivatives of order at most k of u in some compact set of $(x, N) = 0$ has the majorant $O(|\xi|^k)$ for some k. On the other hand

$$|u(N)| = e^{-\operatorname{Im} t(N,N)}, \quad (N, N) > c$$

so that the continuity requirement of the class $H(N)$ means that

$$\operatorname{Im} t(\xi) \ge -C_1 \log(1 + |\xi|) - C_2$$

for some C_1, C_2 and all ξ. What is needed at this point is a lemma which follows from a general result by Tarski and Seidenberg.

LEMMA T. *When $P_m(N) \ne 0$, the minimum $m(r)$ for $|\xi| \le r$ of the imaginary parts of the zeros of $P(tN + \xi) = 0$ is a real algebraic function of r, i.e. there is a polynomial $f(x, y)$ in two variables such that $f(m(r), r) = 0$ for all large r.*

It follows that $\operatorname{Im} t(\xi) \ge - \operatorname{const}$ and the same reasoning at the point $x = -N$ achieves the proof of Lemma 1.

Remark. The sum $\operatorname{Im} \tau_k(\xi)$ of (1) is a linear form of ξ which by the previous lemma is bounded from above . Then it is impossible for one $\operatorname{Im} \tau_k(\xi)$ to go away to $-\infty$. Hence, if $\operatorname{Hyp}(N)$ refers to the values $u(x)$ such that $(x, N) > 0$, we have $\operatorname{Hyp}(-N) = \operatorname{Hyp}(N)$.

Algebraic properties of hyperbolic polynomials

To prove the rest of the Theorem, we need a section on the algebraic properties of hyperbolic polynomials.

LEMMA 2. *If $P \in \operatorname{Hyp}(N)$, then $P_m \in \operatorname{Hyp}(N)$ and all the zeros $\lambda(\xi)$ of*

$$P_m(tN + \xi) = P_m(N) \prod (t + \lambda_k(\xi))$$

are real for real ξ.

PROOF. The imaginary parts of the $\tau_k(\xi)$ of (1) are bounded and, as $t \to \infty$, the $t^{-1}(\tau_k(t\xi))$ tend to the $\lambda_k(\xi)$ in some order. Hence all the $\lambda_k(\xi)$ are real when ξ is real. In other words, $P_m(tN + \xi) \ne 0$ when $\operatorname{Im} t \ne 0$.

DEFINITION. Let $\Gamma(P, N)$ be the set of vectors η such that $P_m(\eta + tN) \neq 0$ for all $t \geq 0$.

Note. This definition means that all $\lambda_k(\eta)$ are positive. Let us now define $c(\eta) \geq 0$ by the requirement that

$$(4) \qquad \qquad \operatorname{Im} s > c(\eta) \implies P(s\eta + \xi) \neq 0$$

for all real ξ. So far we only know that $c(N) < \infty$.

MAIN LEMMA. *If* $\operatorname{Im} t > c(N)$, $\eta \in \Gamma(P, N)$ *and* $\operatorname{Im} s \geq 0$, *then*

$$P(tN + s\eta + \xi) \neq 0$$

for all real ξ.

PROOF. If $w > 0$, the equation $w^{-m} P(w(tN + s\eta) + \xi) = 0$ with $\operatorname{Im} t > c(N)$ has no zeros s on the real axis, and the polynomial involved has the limit

$$P_m(tN + s\eta) = P_m(N) \prod (t + s\lambda_k)$$

as $w \to \infty$. The zeros in s of this limit are all in the lower half-plane since all λ_k are positive. Hence the first polynomial can have no zero s in the upper half-plane. In fact, such a zero must cross the real axis when $w \to \infty$. This proves the lemma.

LEMMA 3. *If* $P \in \operatorname{Hyp}(N)$, *then* $P \in \operatorname{Hyp}(\eta)$ *for all* $\eta \in \Gamma(P, N)$. *A function* $c(\eta)$ *satisfying* (4) *exists,* $c(t\eta) = c(\eta)/t$ *when* $t > 0$ *and* $c(\eta)$ *is bounded on compact subsets of* $\Gamma(P, N)$.

PROOF. By the preceding lemma, $P(tN + s\eta + \xi) \neq 0$ when $t > c(N)$, $\eta \in \Gamma(P, N)$ and $\operatorname{Im} s \geq 0$. For small enough $\varepsilon > 0$, $\eta - \varepsilon N \in \Gamma(P, N)$ and we get

$$P(s\eta + \xi) = P(s\varepsilon N + s(\eta - \varepsilon N) + \xi) \neq 0$$

for all real ξ when $\operatorname{Im} s > C(N)/\varepsilon$. Hence $c(\eta) < c(N)/\varepsilon$. This proves the lemma and proves that the set $\Gamma_1(P, N)$ of all $\eta \in \Gamma(P, N)$ for which $c(\eta) < 1$ is open and has the property that the union of all $s\Gamma_1(P, N)$ for $s > 0$ equals $\Gamma(P, N)$.

The fundamental solution with support in a cone

Let $\eta \in \Gamma_1(P, N)$. Then

$$|P(\xi + i\eta)| \geq |P_m(\eta)|$$

for all real ξ. When $f \in C_0^\infty(R^n)$, consider the distribution

$$(E, f) = (2\pi)^{-n} \int F(\xi + i\eta) d\xi / P(\xi + i\eta), \quad F(\xi + i\eta) = \int e^{(ix, \xi + i\eta)} f(x) dx,$$

where $\eta \in \Gamma_1(P, N)$. Note that

$$(E, P(-D)f) = (2\pi)^{-n} \int F(\xi + i\eta) d\xi = f(0)$$

so that $P(D)E(x) = \delta(x)$, i.e. $E(P, x)$ is a fundamental solution.

By Cauchy's theorem the integral defining E is independent of η as long as $\eta \in \Gamma_1(P, N)$. If $(x, \eta) < 0$ in the support of f, $F(\xi + i\eta)$ tends to 0 uniformly together with all its derivatives if η is replaced by $t\eta$ and t tends to $+\infty$. It follows that $(E, f) = 0$. Hence $E(P, x) = 0$ outside the *propagation cone*

$$K(P < N) = (x; (x, \eta) \geq 0, \quad \text{all } \eta \in \Gamma(P, N)).$$

Since $\Gamma(P, N)$ is open, this cone is proper, i.e. its sections with hyperplanes $(x, \eta) = $ const are compact when $\eta \in \Gamma(P, N)$.

End of the proof of the theorem

If $P(D)u = 0$ and $f \in C_0^\infty$ has compact support, then uf has compact support so that

$$f(x)u(x) = \int E(P, x - y)P(D)u(y)f(y)dy.$$

Since the integrand on the right vanishes outside $x - K(P, N)$, the formula holds when (x, N) is bounded from below on the support of f. In particular, we can take $f(x) = g((x, N))$ where $g(t)$ equals 0 when $t < -c$ and 1 for $t > c$. Then the formula shows that u tends to zero locally uniformly for $(x, N) \geq c$ when u and all its derivatives up to a certain order tend to zero in every compact part of the strip $|(x, N)| < c$. A technical argument, not given here, shows the same assertion when u and all its derivatives up to a certain order tend to zero only in the hyperplane $(x, N) = 0$. Since $\text{Hyp}(N) = \text{Hyp}(-N)$, this completes the proof of the theorem. It follows from the proof that $\text{Hyp}(N) = H(N)$ and that $P \in H(N)$ for some $N \neq 0$ if and only if $P(D)$ has a fundamental solution with support in a proper cone.

Lineality, strong hyperbolicity

The lineality $L(P)$ of a polynomial $Q(\xi)$ is the set of vectors η such that $Q(\xi + t\eta) = Q(\xi)$ for all t and ξ. What this means is that Q only depends on the variables in a complement of $L(P)$. It is not difficult to see that $L(P) = L(P_m)$ when P is hyperbolic. It is more difficult to see that if $P \in \text{Hyp}(N)$, then $P_m + R$ with $\deg R < m$ belongs to $H(N)$ if an only if

$$\frac{R(iN + \xi)}{P_m(iN + \xi)}$$

is a bounded function of ξ. When P is strongly hyperbolic, i.e. when $|\xi|^{m-1} = O(|P(iN + \xi)|)$, then any R of degree $< m$ satisfies this condition.

Support and singular support of a fundamental solution

For strongly hyperbolic operators P with principal part P_m, the singular support of $E(P, N, x)$ is the intersection of the propagation cone and the propagation surface, defined as the dual of the real hypersurface $P_m(\xi) = 0$, generated by $\text{grad} P_m(\xi)$. For the wave operator this is simply the boundary of the propagation cone. A lacuna for the operator P is defined as a region inside the propagation cone but outside the propagation surface where the fundamental solution vanishes. It is a classical fact that the interior of the propagation cone is a lacuna for the wave operator in an even number of variables. For homogeneous and strongly hyperbolic operators Petrovskii (1945) found necessary and sufficient conditions for a point x to be in a lacuna which is stable under small changes of the operator. His condition involves the topology of the complex intersection of the hyperplane $(x, \xi) = 0$ and the hypersurface $P(\xi) = 0$. Petrovskii's condition and the definition of the propagation surface were extended to homogeneous hyperbolic operators, not necessarily strongly hyperbolic, by Atiyah, Bott and Gårding (1970, 1973).

Bibliography

ATIYAH M.F., BOTT R., GÅRDING L.

   1970. *Lacunas for hyperbolic differential operators*, I, Acta Math. **124** (1970), 109–189.

   1973. *Lacunas for hyperbolic differential operators*, II, Acta Math. **131** (1973), 145–206.

COURANT R., HILBERT D.

   1931. *Methoden der Mathematischen Physik*, I, II, Springer, 1931.

GÅRDING L.

   1951. *Linear hyperbolic partial differential equations with constant coefficients*, Acta Math. **85** (1951), 1–62.

HÖRMANDER L.

   1983. *The analysis of linear partial differential operators*, I, II, Grundlehren der Math. Wiss., vols. 256, 257, Springer, 1983.

PETROVSKII I. G.

   1938. *On Cauchy's problem for systems of partial differential equations and non-analytic functions*, Bull. Moscow Univ. Math. and Mech. **1** (1938), no. 7.

   1945. *On the diffusion of waves and lacunas for hyperbolic equations*, Mat. Sb. **17 (59)** (1945), 289–370.

CHAPTER 9

Hypoellipticity

Introduction

Lars Hörmander's by now classical thesis (1955) gave the first general theory of partial differential operators with constant coefficients. Such operators can be written as complex polynomials $P(D)$ in the imaginary gradient $D_k = \partial/i\partial x_k$, where $x = (x_1, \ldots, x_n)$ are n real variables. The arguments of the corresponding characteristic polynomials $P(\xi)$ are n real or complex variables $\xi = (\xi_1, \ldots, \xi_n)$. Let us say for simplicity that an operator $P(D)$ is complete if $P(\xi)$ cannot be expressed as a polynomial in less than n variables.

The maximal domain $A(P)$ of $P(D)$ in an open set Ω was defined as the set of all $f \in L^2(\Omega)$ such that $P(D)f \in L^2(\Omega)$, naturally taken in the weak sense so that the linear functional

$$g \to \int f(x)P(-D)g(x)dx$$

is continuous in $L^2(\Omega)$ when $f \in L^2(\Omega)$.

A main point of Hörmander's thesis is the characterization of operators of local type with the property that $A(\Omega)$ is invariant under multiplication by infinitely differentiable functions with compact supports in Ω. By coincidence, the class of complete operators of local type is also the class of hypoelliptic operators with the property that all local distribution solutions u of $P(D)u(x) = 0$ are infinitely differentiable. At the time of the thesis, the characterization of hypoelliptic operators was a problem posed by Laurent Schwartz, thus unwittingly solved by Hörmander.

Remark. That $P(D)u = 0$ in an open set Ω means of course that the linear form $u(P(-D)g)$ vanishes for all $g \in C_0^\infty(\Omega)$. When Laurent Schwartz taught distribution theory in the late forties, it was forbidden to write linear forms like $u(P(-D)g)$ in symbolic integral form as

$$\int u(x)P(-D)g(x)dx,$$

but soon thereafter, this irresistible and convenient misuse of terminology became widespread.

The purpose of this lecture is to give the simple proof that an operator $P(D)$ is hypoelliptic if and only if

(1) $$P(\xi + i\eta) \neq 0$$

when η is bounded and ξ is large enough. The Tarski-Seidenberg theorem is an essential ingredient of the proof.

A necessary condition

Let Ω be an open, bounded subset of R^n and consider weak solutions $u(x)$ of $P(D)u = 0$ in Ω, i.e. distributions u such that

$$u(P(-D)f) = 0$$

for all $f \in C_0^\infty(\Omega)$. The space S of such solutions which are also in $L^2(\Omega)$ is then obviously a closed subset of $L^2(\Omega)$. By assumption, all element of S are infinitely differentiable. Hence, if Ω contains zero, the map

$$Su \to \operatorname{grad} u(0)$$

is closed and hence continuous so that

$$|\operatorname{grad} u(0)|^2 \leq \operatorname{const} \int_\Omega |u(x)|^2 dx.$$

Inserting exponential solutions,

$$u(x) = e^{i(x,\xi+i\eta)}, \quad P(\xi + i\eta) = 0,$$

it follows that

$$|\xi + i\eta| \leq C_1 e^{C_2|\eta|}$$

for some $C_1, C_2 > 0$. Hence (1) follows. Using the Seidenberg-Tarski theorem, we can also prove a stronger result, namely

LEMMA 1. *If $P(\xi)$ is hypoelliptic, there are real numbers a, b such that $a > 0$ and*

$$|\eta| \leq |\xi|^a \implies |P(\xi + i\eta)| \geq |\xi|^b$$

when $|\xi|$ is large enough.

PROOF. Consider first the real semi-algebraic set

$$t = |\xi|^2, \quad s \geq |\eta|^2, \quad P(\xi + i\eta) = 0.$$

By the Seidenberg-Tarski theorem its projection S' on the (s, t)-plane is a real semi-algebraic set where, according to (1), $s > 0$ must tend to infinity as $t \to \infty$. Hence the function $\varphi(t) = \min s$ for $(s, t) \in S'$ must for large t ultimately lie on a branch of an algebraic curve

$$\varphi(t) = at^b(1 + o(1))$$

where $a, b > 0$ and b is rational. Returning to the variables ξ, η, this implies a stronger version of (1), namely

$(1')$ $$P(\xi + i\eta) = 0 \implies |\eta| \geq \sqrt{a}|\xi|^{b/2}(1 + o(1))$$

for all sufficiently large $|\xi|$. We shall also need to estimate $P(\xi + i\eta)$ in a region as above where it does not vanish. To this end consider a new semialgebraic set T where

$$t = |\xi|^2, \quad |\eta|^{2a_1} \leq t^{a_2}, \quad s \geq |P(\xi + i\eta)|^2$$

where a_1 and a_2 are positive integers chosen so that, by $(1')$, $s \neq 0$ when t is large enough. The projection of T on the (r, s)-plane is then also a semialgebraic set where the infimum σ of s is a piecewise algebraic function of t which does not vanish for large t. Hence it has the form

$$\sigma = a + ct^b(1 + o(1)) \geq 0$$

where a, c are real, $c > 0$, b is rational and t is sufficiently large. It follows that σ is bounded from below or else exceeds some power of t for large t. This proves the lemma.

A sufficient condition

All which remains now is to prove that every hypoelliptic operator $P(D)$ has a fundamental solution $E(x)$,

$$P(D)E(x) = \delta(x),$$

which is infinitely differentiable outside the origin. In fact, then Weyl's lemma as presented in the next lecture (p. 66) shows that every weak solution u of $P(D)u = 0$ is infinitely differentiable. We shall see that every operator satisfying $(1')$ has such a fundamental solution and with this the announced characterization of hypoelliptic operators will be proved.

The construction will succeed if we use the freedom of complex integration in the classical formal expression

$$E(x) = (2\pi)^{-n} \int \frac{e^{i(x,\xi)}}{P(\xi)} d\xi$$

for a fundamental solution. In the first place, let us choose variables ξ so that

$$P(\xi) = c\xi_1^m + \text{lower terms}$$

where $c \neq 0$ and $m = \deg P$. It follows from $(1')$ that there is a number $A > 0$ such that $p(\xi) \neq 0$ when $|\xi| \geq A$. We shall split the formal integral into two parts,

$$E_1(x) = (2\pi)^{-n} \int_{\Gamma_1} \frac{e^{i(x,\zeta)}}{P(\zeta)} d\zeta$$

and

$$E_2(x) = (2\pi)^{-n} \int_{\Gamma_2} \frac{e^{i(x,\zeta)}}{P(\zeta)} d\zeta.$$

Here ζ is put equal to $\xi + i\eta$ in order to mark that some integration takes part in the complex C^n and Γ_1 and Γ_2 are chains whose sum is homologous to R^n in a band around R^n. In Γ_2, the variables $\zeta_2, \ldots, \zeta_n$ are real and their absolute values do not exceed A while ζ_1 runs through a curve in the complex plane from $-A$ to $+A$ which has all the zeros of

$$\zeta_1 \to P(\zeta_1, \zeta_2, \ldots, \zeta_n)$$

on one side. On Γ_1 all variables are real and at least one $|\xi_k|$ exceeds A. It is clear that $E_2(x)$ is an analytic function of x and that $E_1(x)$ is a distribution. When $f \in C_0^\infty$, $E_1 + E_2$ is a distribution $E = E(P, x)$ defined by

$$(2) \qquad (E, f) = (2\pi)^{-n} \int_{\Gamma_1 + \Gamma_2} \frac{F(\zeta)}{P(\zeta)} d\zeta$$

where

$$F(\zeta) = \int e^{i(x,\zeta)} f(x) dx$$

is the Fourier-Laplace transform of f. By the lemma, $1/|P(\xi)|$ does not grow more than a power of $|\xi|$. Hence the integrals converge since

$$|F(\xi + i\eta)| = O((1 + |\xi|)^{-N})$$

for all $N > 0$ when $|\eta|$ is bounded. Moreover, since $P(-D)f$ has the transform $P(\zeta)F(\zeta)$ and since $\Gamma_1 = \Gamma_2$ is homologous to R^n, we have

$$(E, Pf) = (2\pi)^{-n} \int_{R^n} e^{i(x,\zeta)} d\zeta = f(0)$$

so that E is a fundamental solution.

To see that $E(P, x)$ is infinitely differentiable at a point $x_0 \neq 0$, we shall now replace Γ_1 by a chain Γ_1' in C^n whose projection on R^n equals Γ_1 and has the form

$$\xi + i\varphi(\xi)\eta_0.$$

Here η_0 is chosen so that $(x_0, \eta_0) > 0$ while $\varphi(|\xi|) \geq 0$ is continuous and vanishes except for very large ξ where

$$\varphi(\xi) = |\xi|^c$$

and c is chosen so small that Γ_1' is contained in the region described in the lemma. Then, for x close to x_0, the integrand of the integral

$$I(x) = (2\pi)^{-n} \int_{\Gamma_1'} e^{i(x,\zeta)} d\zeta / P(\zeta)$$

is exponentially decreasing so that the integral is an infinitely differentiable function of x. Moreover, if the support of $f \in C^\infty$ is close enough to x_0, we have

$$\int f(x)I(x) = (2\pi)^{-n} \int_{\Gamma_1'} F(\zeta) d\zeta / P(\zeta).$$

Here the integrand is uniformly convergent and unchanged if Γ_1' is modified to Γ_1 by replacing $\varphi(|\xi|)$ by $s\varphi(|\xi|)$ and letting $s > 0$ decrease from 1 to 0. Comparing this with the formula (2) shows that, in fact, $E(x)$ is infinitely differentiable when $x_0 \neq 0$.

Bibliography

HÖRMANDER LARS.
  1955. *On the theory of general partial differential operators*, Acta Math. **94** (1955), 161–248.

CHAPTER 10

Dirichlet's Problem and Gårding's Inequality

Introduction

The purpose of this lecture is to sketch in historical order the various ways of solving Dirichlet's problem and to show that Hermann Weyl's paper *The method of orthogonal projection in potential theory* (1940) made it possible to separate regularity results from the tools of functional analysis and to base the solution of Dirichlet's problem for general elliptic equations on a single result known as Gårding's inequality. For a fuller early history of Dirichlet's problem the reader is referred to Kellogg (1953) and the references there.

A short history of Dirichlet's problem. Weyl's lemma. The method of orthogonal projection

The classical Dirichlet problem can be stated as follows: given an open, bounded set Ω and a continuous function f at the boundary, find a harmonic function u in Ω which assumes the values f at the boundary. For the disk or the ball, the problem has an explicit solution given by the well-known Poisson formula. In his analysis of locally analytic functions on a Riemann surface, Riemann suggested that the solution u of the problem minimizes Dirichlet's integral

$$\int_\Omega (v_x^2 + v_y^2)dxdy$$

for all functions which are equal to f at the boundary. A formal verification is easy, but some years later Weierstrass cast doubt on Riemann's idea by exhibiting variational problems without solutions.

The next step was taken by Carl Neumann who sought a solution u of Dirichlet's problem in the form of a double layer potential on $S = \partial\Omega$ with a continuous density $g(t)$,

$$u(x) = \int_0^L \kappa(x, h(t))g(t)dt.$$

Here $x = (x_1, x_2)$, L is the length of S, t its arc-length parameter, $t \to h(t)$ is the equation of S and

$$\kappa(x, h(t)) = \frac{1}{\pi}D_t \log \frac{1}{|x - h(t)|} = \frac{1}{\pi}(x - h(t), n(t))/|x - h(t)|^2,$$

where the differentiation D_t in the direction of the interior unit normal is carried out in the second half of the formula. It is then clear that $u(x)$ is a harmonic function in Ω. When S has a continuous curvature and x approaches a point $h(s) \in S$, $u(x)$ has the limit

$$f(s) + \int_0^L K(s,t)f(t)dt$$

where the kernel $K(s,t) = \kappa(h(s), h(t))$ is continuous. Hence Dirichlet's problem reduces to an integral equation

$$f(s) + \int_S K(s,t)f(t)dt = g(s)$$

for $f(s)$, or in symbolic form, $f + Kf = g$. When Ω is convex, it turned out that the geometric or Neumann series $(1 - K + K^2 - K^3 + \cdots)g$ converges and produces the solution f. Some twenty years after Neumann, the Swede Ivar Fredholm solved the integral equation above in general. His main result was that the solutions of the homogeneous equation $f + Kf = 0$ form a linear space L of finite dimension and that the inhomogeneous equation $f + Kf = g$ is solvable (necessarily modulo L) when g is subject to $\dim L$ suitably chosen linear conditions. This result, the once famous Fredholm alternative, can also be expressed by saying that the operator $I + K$ acting on continuos functions has a vanishing index, i.e. the dimension of its null space equals the codimension of its range.

Fredholm's work became the starting point for a long period of intense interest in integral equations. One of the fruits was Hilbert's spectral theorem for bounded selfadjoint operators on Hilbert space and von Neumann's extension to unbounded operators. Both theorems are now anonymous parts of a general mathematical education.

In the meantime, Dirichlet's problem had been solved in two new ways, by Poincaré who made repeated use of the solution for disks (the sweeping out method), and by Hilbert who rehabilitated Dirichlet's principle by taking suitable minimizing sequences. Finally, there was a very elementary solution by Perron using superharmonic and subharmonic functions.

The last milestone in this parade of solutions to Dirichlet's problem is Hermann Weyl's paper quoted above. The main point of this paper is Weyl's lemma: if u is locally square integrable in a region $\Omega \subset R^n$ and

$$\int u(x)\Delta f(x)dx = 0, \quad x = (x_1, \ldots, x_n) \in R^n, \quad \Delta = -\sum(\partial/\partial x_k)^2$$

for all smooth functions f with compact supports in Ω, then $u(x)$ is almost everywhere equal to a harmonic function, in particular locally analytic in the variables x. We shall give a simple proof which extends the lemma to distributions.

The standard fundamental solution $E(x) = c_n|x|^{2-n}$ of Δ when $n > 2$ and the corresponding logarithm when $n = 2$ is analytic for $x \neq 0$. Let $f(x) \in C_0^\infty(\Omega)$ be equal to 1 in an open subset ω and consider the integral

$$(1) \quad v(x) = \int u(y)w(x,y)dy, \quad w(x,y) = \Delta(1 - f(y))E(x - y), \quad \Delta = \Delta_y,$$

for $x \in \omega$. Notice first that $\Delta E(x - y) = 0$ when $x \neq y$. Hence the functions $y \to w(x,y)$ vanish when $f = 0$ or $f = 1$ and must have compact supports in Ω. Moreover, $w(x,y)$ is infinitely differentiable when $x \in \omega$. Hence the integral $v(x)$ is infinitely differentiable. If $g \in C_0^\infty(\omega)$, we can form the integral of $v(x)g(x)$, getting

$$\int v(x)g(x)dx = \int u(y)\Delta(1 - f(y))h(y)dy, \quad h(y) = \int E(x - y)g(x)dx.$$

Since E is a fundamental solution, the right side can be written as

$$\int u(y)g(y)dy - \int u(y)\Delta f(y)h(y)dy$$

where $fh \in C_0^\infty(\Omega)$ so that the second term vanishes. Hence $u = v$ almost every-where in ω and the lemma follows. If the relevant integrals are taken as the values of a distribution u, the same proof shows that a distribution u such that $\Delta u = 0$ is represented by a harmonic function.

Remark. Given certain conditions, the formula (1) extends to other cases. We may replace Δ by a differential operator $P(x, D)$ with smooth coefficients which has a fundamental solution $E(x, y)$,

$$P(y, D_y)E(x, y) = \delta(x - y),$$

which is infinitely differentiable when $x \neq y$. Just replace $w(x, y)$ by $P(y, D_y)(1 - f(y))E(x, y)$ and the proof shows that a distribution u such that $u(Pf) = 0$ when f is infinitely differentiable with compact support is represented by a smooth function.

The remark applies to elliptic operators with smooth coefficients, but Weyl's lemma has many other forms. All of them state in some way or other that there are differential operators, in particular the elliptic ones, whose weak solutions have strong regularity properties. The main impact of Weyl's lemma has been to separate boundary problems for certain differential operators into two parts, first some kind of existence and then an analysis of the regularity properties of the solutions. We shall demonstrate this with Weyl's treatment of Dirichlet's problem.

In fact, assume that Ω is bounded and let H_0^1 be the closure in the square norm of the inner product

$$(2) \qquad (u, v) = \int_\Omega \sum u_{x_k} \bar{v}_{x_k} dx$$

of the space C_0^1 of once continuously differentiable functions with compact supports in Ω. It is a classical fact that the imbedding $L^2(\Omega) \to H_0^1$ is completely continuous and it is not difficult to verify that all elements of H_0^1 vanish at the boundary of Ω in a certain sense.

These regularity questions left aside, let $w \in H^1$ be such that $(w, w) < \infty$. Then $v \to (v, w)$ is a linear functional in H_0^1 and must have the form (v, w') where $w' \in H_0^1$. Hence the function $u = w - w'$ solves Dirichlet's problem in our new setup: it equals w on the boundary and, by Weyl's lemma, it is a harmonic function in Ω since $(w - w', v) = -\int(w - w')\Delta \bar{v}\, dx = 0$ for all smooth functions v with compact supports. With equal ease we can construct a Green's operator $G : L^2(\Omega) \to H_0^1$ with the property that $\Delta Gv = v$ at least in the weak sense. This operator is defined by the equality $\int v\bar{w}\, dxdy = (Gv, w)$ where $v \in L^2$ and $w, Gv \in H_0^1$ and the left side is an antilinear functional of w. Writing out this equality in a weak form as

$$(v, w) = (Gv, \Delta w)$$

for $w \in C_0^2(\Omega)$ proves that $\Delta Gv = v$ in the weak sense.

We may even permit imaginary parts and lower order terms. In fact, suppose that

$$M = i \sum \partial_{x_j} a_{jk}(x)\partial_{x_k}, \quad L = \sum b_j \partial_{x_j} + b_0$$

with bounded and regular coefficients, those of M being real. Then we have by integrations by parts,

$$\int (\Delta + M + L)u(x)\bar{v}(x)dx = ((1 + iA + B)u, v)$$

where $A : H^1 \to H^1_0$ is a bounded operator, selfadjoint on H^1_0, and $B : H^1 \to H^1_0$ is a bounded operator, compact on H^1_0. Hence the operator $1 + iA + tB$ is invertible except for a discrete set of eigenvalues t tending to ∞. Outside these values it is invertible and hence, given $w \in H^1$, there is a $w' \in H^1_0$ such that $(1 + iA + tB)(w - w') = 0$. It follows that $u = w - w'$ solves Dirichlet's problem for $\Delta + M + tL$ with the same boundary values on $\partial\Omega$ as w.

Gårding's inequality

The origin of Gårding's inequality was an effort to solve Dirichlet's problem for linear elliptic operators of high even order. Together with the simple functional analysis, sketched above, the inequality achieved its aim. The problem is to find a suitable extension of the inner product (2). Let

$$P(x, D) = \sum a_\alpha(x) D^\alpha$$

with principal part

$$P_{2m}(D) = \sum_{|\alpha|=2m} a_\alpha(x) D^\alpha$$

have the property that

(3) $$\operatorname{Re} P_{2m}(x, \xi) \geq c|\xi|^{2m}$$

for some $c > 0$ and x in the closure of some open set Ω. It is further assumed that the coefficients $a_\alpha(x)$ have continuous and bounded derivatives of order $\leq [|\alpha| - m]_+$.

THEOREM. *Under these circumstances,*

(4) $$\operatorname{Re}(Pu, u) \geq (c - \varepsilon)|u|^2_m - b_\varepsilon \|u\|^2_0$$

for all $u \in C^m_0(\Omega)$, every $\varepsilon > 0$ and some $b_\varepsilon < \infty$. Here

$$|u|^2_k = \int \sum_{|\alpha|=k} |D^\alpha u(x)|^2 dx.$$

Remark 0. Under suitable assumptions, the inequality extends to operators P which are square matrices.

Remark 1. Since $u \in C^m_0$ and $a_\alpha(x)$ has bounded derivatives of order $\leq [|\alpha| - m]_+$, the integral

$$(Pu, u) = \int \sum_{|\alpha| \leq 2m} a_\alpha(x) D^\alpha u(x)\bar{u}(x)dx$$

can be written in the form

(5) $$(Au, u) = \int \sum a_{\alpha\beta}(x) D^\alpha u(x) D^\beta \bar{u}(x)dx, \quad |\alpha| \leq m, |\beta| \leq m,$$

with bounded and continuous coefficients.

Remark 2. It suffices to prove (4) in the form

$$(4') \qquad \operatorname{Re}(Au, u) \geq (c - \varepsilon)|u|_m^2 + O(\|u\|_m \|u\|_{m-1})$$

where $\|u\|_k = |u|_k + |u|_{k-1} + \cdots + |u|_0$. In fact,

$$\|u\|_m \|u\|_{m-1} \leq \varepsilon \|u\|_m^2 + \frac{1}{4\varepsilon} \|u\|_{m-1}^2$$

for every $\varepsilon > 0$ and

$$\|u\|_{m-1} \leq \varepsilon |u|_m + b_\varepsilon |u|$$

where ε and b_ε have their generic significance.

PROOF. Let us first assume that $P = P(D)$ has constant coefficients. Then by Parseval's formula,

$$(Pu, u) = (2\pi)^{-n} \int P(\xi) |\hat{u}(\xi)|^2 d\xi$$

where $\hat{u}$ is the Fourier transform of u. According to (3)

$$\operatorname{Re} P(\xi) \geq c|\xi|^{2m} - c_1(|\xi|^{2(m-1)} + 1)$$

where the right side is minorized by

$$(c - \varepsilon)|\xi|^{2m} - b_\varepsilon$$

for every $\varepsilon > 0$ and some b_ε. Hence, by Parseval's formula again, the theorem follows in this case.

The proof for variable coefficients depends on a partition of unity,

$$1 = \sum f_k(x)^2$$

where the f_k are infinitely differentiable and so small that

$$|\operatorname{Re} P_{2m}(x, \xi) - \operatorname{Re} P_{2m}(y, \xi)| \leq \frac{\varepsilon}{4} |\xi|^{2m}, \quad x, y \in \operatorname{supp} f_k$$

for all k. Such a partition exists. We then have

$$(Au, u) = \sum (f_k^2 Au, u) + O(\|u\|_m \|u\|_{m-1}).$$

Now, moving the two f_k from this position to $f_k u$, just involves products of derivatives $D^\alpha u D^\beta \bar{u}$ with $|\alpha|, |\beta| \leq m$ and $|\alpha + \beta| < 2m$ and hence changes the sum on the right by $O(\|u\|_m \|u\|_{m-1})$ and hence

$$\operatorname{Re}(Au, u) \geq \sum \operatorname{Re}(Af_k u, f_k u) + O(\|u\|_m \|u\|_{m-1}).$$

Here, by a preceding formula, every term in the sum is minorized by

$$(c - \frac{\varepsilon}{2})|f_k u|_m^2 + O(\|u\|_m \|u\|_{m-1})$$

and this term differs from

$$(c - \frac{3\varepsilon}{4}) \int \sum_{|\alpha|=m} |f_k^2 D^\alpha u(x)|^2 dx = (c - \frac{3\varepsilon}{4})|u|_m^2$$

by the same error term as before. Hence

$$\operatorname{Re}(Au, u) \geq (c - \frac{3\varepsilon}{4})|u|_m^2 + O(\|u\|_m \|u\|_{m-1}).$$

Here the error term is majorized by

$$\frac{\varepsilon}{4}|u|_m^2 + b|u|^2$$

when b is large enough and hence the result follows.

It is clear from the end of the first section how to use the inequality in order to solve Dirichlet's problem. The crucial form (3) is then replaced by $\mathrm{Re}((A+t)u, u)$ with t large positive and the corresponding bilinear form.

Final remarks

The original Gårding's inequality seems to have been used for the first time by Leray in his 1954 lectures for an existence proof for Cauchy's problem for strongly hyperbolic systems. It was used again by Gårding (1956) for a proof of the same result using only functional analysis.

In retrospective, the inequality appears as the first and simplest result in the theory of boundary problems for high order elliptic equations and systems which was developed in the fifties and sixties (see for instance Hörmander (1963), Chapter X).

The fact that the inequality is a well-known result is not easy to explain. One reason is perhaps that it extends easily to pseudodifferential operators and in this way it has been quoted many times. Finally there is a sharp form of the inequality due to Hörmander (see Hörmander (1984)) which says that if the symbol $P(x, \xi)$ has order $2m$ and $\mathrm{Re}\, P(x, \xi) \geq 0$. Then

$$\mathrm{Re}(P(x, D)u, u) \geq -\,\mathrm{const}\,\|u\|_{m-1/2}^2.$$

Notice that $2m$ here may be any real number and that the error term has the order of the terms thrown away in the proof above. A still sharper form of the inequality is due to Melin (1971).

Bibliography

GÅRDING L.

    1953. *Dirichlet's problem for linear elliptic differential operators*, Math. Scand. **1** (1953), 55–72.

    1956. *Solution directe du problème de Cauchy pour les éqations hyperboliques*, Coll. CNRS Nancy, 1956.

HÖRMANDER L.

    1963. *Linear partial differential operators*, Grundlehren vol. 116, Springer, 1963.

    1964. *The analysis of linear partial differential operators*, III, Grundlehren vol. 274, Springer, 1984.

LERAY J.

    1954. *Hyperbolic differential equations*, Inst. Adv. Study, Prinecton, NJ, 1954.

KELLOG O. D.

    1953. *Foundations of potential theory*, Dover, 1953, original 1929.

MELIN A.

    1971. *Lower bounds for pseudodifferential operators*, Arkiv Mat. **9** (1971), 117–140.

WEYL H.

    1943. *The method of orthogonal projection in potential theory*, Duke Math. J. **7** (1940,), 411–444.

CHAPTER 11

A Sharp Form of Gårding's Inequality

The calculus of pseudodifferential operators

The aim of this chapter is to describe and prove an extension of Gårding's inequality to pseudodifferential operators. It therefore starts with a short and self-contained description of the calculus of pseudodifferential operators.

The symbols $P(x,\xi)$ of differential operators $P(x,D)$ are polynomials in the variable $\xi \in R^n$. Pseudodifferential operators $a(x,D)$ have symbols $a(x,\xi)$ which are infinitely differentiable in all variables and, in the simplest case,

$$\partial_x^\alpha \partial_\xi^\beta a(x,\xi) \leq C_{\alpha,\beta}(1+|\xi|)^{m-|\beta|}$$

for real m and large ξ. These inequalities define a space S^m of symbols where symbols which satisfy the above for all m are considered negligible. When $a_k(x,\xi) \in S^{m-k}$, and $\varphi_k(\xi)$ are smooth functions such that $1 - \varphi_k(\xi)$ have compact supports tending to infinity sufficiently fast, it is easy to see that the series $\sum \varphi_k(\xi)a_k(\xi)$ converges to an element of S^m. Sums $\sum a_k(x,\xi)$ with $a \in S^k$ and $k \to -\infty$ will be taken in this sense below.

A pseudodifferential operator $a(x,D)$ with the symbol $a(x,\xi)$ operates on the space $\mathscr{S}$ of tempered functions via the formula

$$a(x,D)u(x) = (2\pi)^{-n} \int e^{i(x,\xi)} a(x,\xi)\hat{u}(\xi)d\xi$$

where $\hat{u}(\xi)$ is the Fourier transform of u,

$$\mathcal{F}u(\xi) = \int e^{-i(x,\xi)} u(x)dx.$$

Since a product $x^\alpha a(x,D)u(x)$ equals an integral above where D_ξ^α operates on $a(x,\xi)\hat{u}(\xi)$ it is easy to convince oneself that $a(x,D)$ is a continuous map $\mathscr{S} \to \mathscr{S}$. It is less easy to show that $\|a(x,D)u\| \leq \text{const} \|u\|_m$ if $a \in S^m$ where

$$\|u\|_m^2 = \int |\hat{u}(\xi)|^2 (1+|\xi|^2)^m d\xi$$

for any real m.

The product $a(x,D)b(x,D)$ of two pseudodifferential operators with symbols $a(x,\xi) \in S^m$ and $b(x,\xi) \in S^p$ is again a pseudodifferential operator $c(x,D) \in S^{m+p}$ with the symbol

(1) $$c(x,\xi) = e^{i(D_y,D_\eta)} a(x,\eta)b(y,\xi)|_{y=x,\eta=\xi},$$

extending a corresponding formula for diffferential operators. The proof is long and will not be given here.

It is clear that a pseudodifferential operator $a(x, D)$ has the distribution kernel

$$K(x, y) = (2\pi)^{-n} \int e^{i(\xi, x-y)} a(x, \xi) d\xi.$$

Hence the adjoint $a^*(x, D)$ of $a(x, D)$ with respect to the duality

$$(u, v) = \int u(x) \bar{v}(x) dx$$

has the kernel

$$K^*(x, y) = \bar{K}(y, x) = (2\pi)^{-n} \int e^{i(x-y, \eta)} \overline{a(y, \eta)} d\eta.$$

Some calculations not given here show that this is the kernel of a pseudodifferential operator $b(x, D)$ with the symbol

$$(2) \qquad b(x, \xi) = e^{i(D_x, D_\xi)} \overline{a(x, \xi)} = \sum i^{|\alpha|} D_\xi^\alpha D_x^\alpha \overline{a(x, \xi)}$$

extending the formula for differential operators. Note that the terms of this formal expansion belong to S^{m-k} for $k = 0, 1, 2, \ldots$ and that the dominating term $\overline{a(x, \xi)}$ is obtained for $\alpha = 0$.

Pseudodifferential operators and an extension of Gårding's inequality

The theory of peudodifferential operators sheds new light on Gårding's inequality which, in its original form, says that if $a(x, D) = \sum a_\alpha(x) D^\alpha$ is a differential operator of even order $2m$ such that $a_\alpha(x)$ has $[|\alpha| - m]_+$ bounded derivatives and if

$$\operatorname{Re} a(x, \xi) \geq c|\xi|^{2m}$$

for some $c > 0$ and sufficiently large ξ, then

$$\operatorname{Re}(a(x, D)u, u) \geq (c - \varepsilon)\|u\|_m^2 - b_\varepsilon \|u\|_0^2$$

for all $\varepsilon > 0$ and constant $b_\varepsilon > 0$. It follows from the proof that nothing is lost if $\|u\|_0$ is replaced by $\|u\|_{m-1/2}$. In a sharp version for pseudodifferential operators due to Hörmander (1966), m may be any real number and the inequality holds with $\varepsilon = 0$ and $b_0 < \infty$. In other words: if $a \in S^{2m}$ and $\operatorname{Re} a(x, \xi) \geq 0$, then

$$(3) \qquad \operatorname{Re}(a(x, D)u, u) \geq - \operatorname{const} \|u\|_{m-1/2}^2.$$

Note that if $\operatorname{Re} a(x, \xi) \geq 0$ is the square of a symbol $b(x, \xi) \in S^m$, there is a simple proof. In fact, then the symbol of $b^*(x, D)b(x, D)$ differs from $\operatorname{Re} a(x, D)$ by an operator $c(x, D)$ of degree at most $2m - 1$ and hence

$$\operatorname{Re}(a(x, D)u, u) = (b(x, D)u, b(x, D)u) + (c(x, D)u, u)$$

so that (3) follows by the properties of pseudodifferential operators.

In this lecture I shall go through Hörmander's proof (III, Chapter 18, pp. 76–78) in a leisurely way.

A smoothing symbol

LEMMA 1. *If $F(x,\xi)$ is a symbol and*

$$(F(x,D)u,u) \geq 0$$

for all $u \in \mathscr{S}$, then also

$$(F((x-y)t),((D-\eta)/t)u,u) \geq 0$$

for all x,y and $t > 0$.

PROOF. If $v(x) = t^{-n/2}u(x/t)$, $t > 0$, then

$$(F(x,D)v,v) = (2\pi)^{-n}\int e^{i(x-y,\xi)}F(x,\xi)t^{-n}u(y/t)\overline{u(x/t)}dx\,dy\,d\xi.$$

Replacing x,y,ξ by $xt,yt,\xi/t$ shows that

$$(F(tx,D/t)u,u) \geq 0$$

for all $u \in \mathscr{S}$ and $t > 0$. Replacing $u(x)$ by $e^{i(\eta,x)}u(x+y)$ then proves the lemma.

LEMMA 2. *There is a symbol $F(x,\xi) \in \mathscr{S}(R^{2n})$, positive and even in (x,ξ) with total integral 1, such that*

$$(F(x,D)u,u) \geq 0$$

for all $u \in \mathscr{S}$.

PROOF. Let $g(x,\xi) \in C_0^\infty(R^{2n})$. Then $g(x,D)$ has the kernel

$$K(x,y) = (2\pi)^{-n}\int e^{i(\xi,x-y)}g(x,\xi)d\xi$$

from which it follows that

$$\int K(x,x)dx = (2\pi)^{-n}\int g(x,\xi)dxd\xi, \quad \int |K(x,y)|^2dy = \int |g(x,\xi)|^2d\xi.$$

Applying this to the operator $F(x,D) = g^*(x,D)g(x,D)$ with the kernel $\int K(x,y)\overline{K(z,y)}dy$ we may choose $\int F(x,\xi)dxd\xi = \int |g(x,\xi)|^2dxd\xi = 1$, and the formulas (1) and (2) show that $F(x,\xi)$ is even when $g(x,\xi)$ is.

The sharp inequality

Let $a(x,D) \in S^{2m+1}$, with the exponent for convenience, be a pseudodifferential operator whose symbol $a(x,\xi)$ is ≥ 0. The idea of the proof is to use the smoothing operator above to prove that

$$h(x,\xi) = \int F((x-y)/q(\eta),(\xi-\eta)q(\eta))a(y,\eta)dy\,d\eta$$

differs from $a(x,\xi)$ by an element in S^{2m} for some choice of $q(\eta) > 0$. In fact, then

$$0 \leq (\operatorname{Re}h(x,D)u,u) = (\operatorname{Re}a(x,D)u,u) + (r(x,D)u,u)$$

where $r(x, \xi) \in S^{2m}$ and it follows that

$$\mathrm{Re}(a(x, D)u, u) \geq -\,\mathrm{const}\,\|u\|_m^2,$$

i.e. the sharp inequality.

The proper choice of q is $q(\eta) = \sqrt{(1 + |\eta|)}$. This is motivated by the proof. To begin with, we shall prove

LEMMA 3. *The integral*

(4) $$\int_{|\xi - \eta| \geq (1 + |\xi|)/2} F((x - y)q(\eta), (\xi - \eta)/q(\eta))(a(x, \xi) - a(y, \eta))\,dy\,d\eta$$

has the majorant $\mathrm{const}(1 + |\xi|)^{-N}$ *for all* $N > 0$.

PROOF. The restriction $|\xi - \eta| \geq (1 + |\xi|)/2$ of the region of integration means that $1 + |\eta| \leq 1 + |\xi| - |\xi - \eta| \leq (1 + |\xi|)/2$ and hence that

(5) $$1 + |\xi - \eta| \geq \frac{1}{\sqrt{2}}\sqrt{(1 + |\xi|)(1 + |\eta|)} = \frac{1}{\sqrt{2}}q(\eta)q(\xi).$$

Since $F \in \mathscr{S}(R^{2n})$,

$$|F((x - y)q(\eta), (\xi - \eta)/q(\eta))| \leq \mathrm{const}(1 + |x - y|q(\eta))^{-N}(1 + |\xi - \eta|/q(\eta))^{-3N}$$

for all $N > 0$. Also, $|a(y, \eta) - a(x, \xi)| = O((1 + |\eta|)^{2m+1} + (1 + |\xi|)^{2m+1})$. Hence an integration with respect to y in (4) when $N > n$ leaves a bound

$$\mathrm{const}\,q(\eta)^{-N}(1 + |\xi - \eta|/q(\eta))^{-3N}((1 + |\eta|)^{2m+1} + (1 + |\xi|)^{2m+1})$$

to be integrated with respect to η. Here we can use (5) to split off

$$(1 + |\xi - \eta|/q(\eta))^{-2N} \leq (1 + |\xi|)^{-N}.$$

Further,

$$q(\eta)^N(1 + |\xi - \eta|/q(\eta))^{-N} \leq (1 + |\xi - \eta|)^{-N}$$

so that an integration with respect to η leaves a final bound $|\xi|^{2m+1-N}$ and this achieves the proof.

To complete the proof of the sharp inequality we must show that the integral

(6) $$\int F((x - y)q(\eta), (\xi - \eta)/q(\eta))(a(y, \eta) - a(x, \xi))\,dy\,d\eta$$

has the majorant $O((1 + |\xi|)^{2m})$. By Lemma 3 we may without loss of generality restrict the integration by the condition that

$$|\xi - \eta| \leq (1 + |\xi|)/2,$$

which means that $\frac{1}{2}(1 + |\xi|) \leq (1 + |\eta|) \leq 2(1 + |\xi|)$. In other words, $q(\xi)$ and $q(\eta)$ are equivalent.

To proceed further, let us change variables so that

$$y' = (x - y)q(\eta), \quad \eta' = (\xi - \eta)/q(\eta).$$

The integral to be estimated is then

(7) $$\int F(y', \eta')(a(x - y'/q(\eta), \xi - \eta'q(\eta)) - a(x, \xi))\,dy'\,d\eta'$$

since $dy'q(\eta)d\eta'/q(\eta) = dy'd\eta'$. Let us now expand the parentheses of (7) according to Taylor's formula up to the second order,

$$(8) \qquad -a_x y'/q - a_\xi \eta' q + \frac{1}{2}a_{xx}(y'/q)(y'/q) + a_{x\xi}y'\eta' + \frac{1}{2}a_{\xi\xi}(\eta'q)(\eta'q) + R.$$

Here $a = a(x,\xi)$, $q = q(\eta)$ and $a_x y'$ means $(\mathrm{grad}_x\, a(x,\xi), y')$ and similarly for the other terms. Since $\int F(y',\eta')d\eta'$ and $\int F(y',\eta')dy'$ are even functions, the first two terms disappear in the integration. Moreover, since $q(\eta) \sim q(\xi)$ and differentiation with respect to ξ decreases the order of $a(x,\xi)$, by one unit, all the coefficients of the second order terms have the majorant $(1 + |\xi|)^{2m}$ which subsists after the integration. It remains to estimate the remainder. Obvious estimates of the third order derivatives give the following estimate:

$$O(r(\xi)^{2m+1})(|y'/q|^3 + |y'|^2|\eta'|/r(\xi)q(\xi) + |y'||\eta'|^2 q(\xi)/r^2(\xi) + |\eta'|^3 q(\xi)^3/r^3(\xi))$$

where $r(\xi) = 1 + |\xi|$ and we can put $q = q(\xi)$. All terms have the same order $O((1 + |\xi|)^{2m-1/2})$ in ξ. Since F has moments of all orders, the proof of the sharp inequality is finished.

Note. The previous estimates also show the part played by the factor $q(\xi)$. It puts all terms of (8) of the same order on an equal footing with respect to the order of growth in ξ. When $F(x,\xi)$ is chosen to be even, the terms of first order disappear in the integration. This artifice together with the construction of the smoothing function F are the basic ingredients of the proof.

Bibliography

HÖRMANDER L.

1966. *Pseudodifferential operators and hypoelliptic equations*, Ann. Math. **83** (1966), 133–183.

1987. *The analysis of linear partial differential operators*, III, Springer, 1987.

CHAPTER 12

The Impact of Distributions in Analysis

Introduction

The book *Théorie des distributions* by Laurent Schwartz (1951), now one of the non-read classics of mathematics, has transformed many branches of analysis, and the theory is now familiar to every student who ever took an advanced mathematics course in analysis. In this chapter, I will try to describe the essential features of the theory, how it was gradually generally accepted, how if found a niche for itself in analysis and how it made microlocal theory possible. I see my text as a personal report from a bystander who has witnessed a fascinating chapter in the development of analysis.

The theory and its reception

The title of Schwartz's first publication (1945), *Generalization of the notion of a function, of differentiation, of Fourier transformation and applications to mathematics and physics*, sums up the aim of the theory. The main idea is that integrable functions $f(x)$ of several variables $x = (x_1, \ldots, x_n)$ give rise to linear functionals

$$\varphi(x) \to T(\varphi) = \int f(x)\varphi(x)dx$$

on the space C_0^∞ of functions $\varphi(x)$ which are infinitely differentiable and have compact supports, and that such linear functionals, called distributions, are much more general than ordinary functions but still can be localized and possess supports just as functions. This is achieved by defining the product of a distribution T and a function $g \in C_0^\infty$ by the formula

$$gT(\varphi) = T(g\varphi).$$

When $gT = 0$ for all g supported in some open set A, T is said to vanish in A. This notion leads readily to the notion of support for distributions and the space of distributions with compact supports. The choice of C_0^∞ as the space of what was later called test functions made it possible to differentiate distributions any number of times:

$$\partial^\alpha T(\varphi) = T((-\partial)^\alpha \varphi).$$

Here $\partial^\alpha = \partial_1^{\alpha_1} \cdots \partial_n^{\alpha_n}$ with $\partial_k = \partial/\partial x_k$. Schwartz's notation $\mathscr{D}$ for the space C_0^∞ of test functions, $\mathscr{D}'$ for the space of distributions and $\mathscr{E}'$ for the subspace of compactly supported distributions was destined to a long life in the mathematical literature.

All these ideas were not entirely new, but when they appeared earlier, for instance in Sobolev (1936), it was almost always in connection with specific problems. In Sobolev's case it was Cauchy's problem for second order hyperbolic equations in an odd number of space variables, and the idea of distribution solutions was used only for a uniqueness theorem. So-called weak solutions were considered by Weyl (1940) who proved that a weakly harmonic function is harmonic, and by Friedrichs (1944) who proved that integrable, weak solutions of first order linear systems are, in a sense, also strong provided the principal coefficients are continuously differentiable and the others continuous. Schwartz's broad attack, his radical use of infinitely differentiable functions and his conviction that distributions would be useful almost everywhere in analysis made the difference.

One of the prizes of the new theory was an attractive definition of the notion of fundamental solution T_y of a partial differential operator $P(x, D) = \sum a_\alpha(x) D^\alpha$ with infinitely differentiable coefficients and adjoint $P'(x, D) = \sum (-D)^\alpha a_\alpha(x)$. It is required that $T_y(P'\varphi) = \varphi(y)$. In a later and more free notation, initially forbidden by Schwartz, this reads as

$$P(x, D)E(x, y) = \delta(x - y),$$

where T_y is now written as a function $E(x, y)$.

In the beginning the continuity of distributions T was described by Schwartz as follows: $T(\varphi)$ tends to zero when φ is supported in a fixed compact set and all its derivatives tend to zero uniformly. But a definition leading to suitable seminorms[1], for a linear topology is almost as easy: for every compact set K there is a constant C and an integer k such that

$$|T(\varphi)| \leq C \sum_{|\alpha| \leq k} \sup |\partial^\alpha \varphi(x)|,$$

for every $\varphi \in C_0^\infty(K)$. This definition, F. Riesz's formula for linear functionals of continuous functions as measures and the Hahn-Banach theorem prove at once a structure theorem: every distribution is locally a sum $\sum \partial^\alpha M_\alpha$ of derivatives of measures M_α,

$$M_\alpha(\varphi) = \int \varphi(x) d\mu_\alpha(x).$$

When the support of T is a point y, it follows as a special case that T is a finite sum of derivatives of the unit measure at y, denoted by the Dirac distribution $\delta(x - y)$.

Translations $L_y\varphi(x) = \varphi(x - y)$ of test functions carry over to distributions, $L_yT(\varphi) = T(L_{-y}\varphi)$ is an infinitely differentiable function of y and the convolution of a distribution and a test function ψ is an infinitely differentiable function

$$T * \psi(\varphi) = T(\check{\psi} * \varphi), \quad \check{\psi}(x) = \psi(-x),$$

which tends weakly to T when letting $\psi \geq 0$ tend to the Dirac measure, i.e. $\operatorname{supp} \psi$ tends to 0 while $\int \psi(x) dx = 1$. This possibility of regularizing a distribution through weak approximation by infinitely differentiable functions is important in practice.

Schwartz's treatment of the Fourier transform is based on the fact that the classical Fourier transformation $\mathcal{F}$,

$$\mathcal{F}\varphi(t) = \int e^{-i(x,t)} \varphi(x) dx,$$

---

[1] $\sum \sup \int a_\alpha(x) |D^\alpha \varphi(x)| dx$ with continuous coefficients $a_\alpha(x) \geq 0$, vanishing on every compact set for sufficiently large $|\alpha|$.

is an isomorphism of the space $\mathscr{S}$ of tempered functions all of whose derivatives have the bound $O((1+|x|)^{-k})$ for all integers k. The appropriate seminorms make $\mathscr{S}$ a Fréchet space. The Fourier transform then operates on the dual space $\mathscr{S}'$ of tempered distributions by the formula

$$\mathcal{F}T(\varphi) = T(\mathcal{F}\varphi).$$

This simple idea of letting the Fourier transform operate on tempered distributions was superior to all earlier or simultaneous efforts by Wiener (1926), Bochner (1932) and Carleman (1944) to extend the Fourier transform beyond integrability. Moreover, the space $\mathscr{S}$ is obviously a module for the space $\mathscr{O}$ of infinitely differentiable functions f of moderate growth, defined by the property that every derivative of f has at most polynomial growth. Hence $\mathscr{S}'$ has the same property and this makes it possible to define operators $f(D)$, $D = d/idx$ on tempered distributions as follows:

$$f(D) = \mathcal{F}^{-1}f(\cdot)\mathcal{F}.$$

Conceptually, this was a giant step from earlier efforts to define operators of the form $f(D)$ when f is not a polynomial by, for instance, Heaviside (1893), Wiener (1926) and Bochner (1932). But the most important legacy of the theory of distributions is perhaps that it freed Fourier analysis from the prison of absolute and square integrability where it was kept before the war by the classical treatments, for instance Titchmarsh's book (1937) on Fourier integrals and Zygmund's *Trigonomerical series* (1935). Schwartz's extension (1952) of the Paley-Wiener theorem to tempered distributions with compact or conical supports has extended the scope of the Laplace transform.

Schwartz's book (1951) is concerned with the theory as summed up above and the precise definitions of the linear spaces involved. In line of applications Dirac's δ-function is seen as a measure and it is proved for instance that a distribution which is non-negative for non-negative functions is a non-negative measure, that harmonic distributions are actually real analytic functions, a result which probably extends to elliptic systems with analytic coefficients, and that the solutions of hyperbolic equations of order two, as computed by Hadamard and M. Riesz, are actually distributions, and so on. These applications do not measure up to the ambitions of the theory and one gets the impression that the author's heart is with the linear topological spaces rather than with the problems of analysis. Nevertheless the theory of distributions was a great step forward in the applications of abstract linear theory to analysis. The previous efforts, some in Banach's book (1932) and others scattered in the literature, seem rather misdirected in comparison.

The initial success of the theory was due to an enthusiastic forward marketing and to Laurent Schwartz's conviction of the importance of distributions and the fiery lectures which he gave in several European countries after the war. I met him the first time at such a lecture in Lund in the fall of 1948. Like many in his audiences I will forever remember his powerful rendering and insistent French intonation of one of his first sentences: *Les fonctions indéfiniment differentiables et nulles en dehors d'un compact.*

In 1946–47, I spent a year at the mathematics department of Princeton University and then witnessed the introduction of the theory of distributions at this Mecca of mathematics, where the principal subjects at the time were topology and algebra. Claude Chevalley gave an introductory colloquium talk followed by a series of lectures by a French visitor, Jean Delsarte from Nancy. There were two persons in his audience and I was one of them. At the time these lectures helped me formulate an intrinsic notion of hyperbolicity for partial differential operators with constant coefficients. But in the finished paper[2] I found no use for distributions.

At the 1950 International Congress in Boston, a Fields medal went to Laurent Schwartz for the theory of distributions. This was perhaps the first time and maybe also the last time that this prize was given for a piece of soft mathematics and not for the solution of some known difficult problem. But the prize committee had shown an impressive foresight.

At the time the theory of distributions got a rather lukewarm and sometimes even hostile reception among mathematicians. Analysts of an older school could joke that "Your distributions may be all right, but you are only really happy when you find a function." Bochner's review (1953) of *Théorie des distributions* has a heavy, sarcastic ending: "We have recounted all this with a view suggesting that it would not be easy to decide what general innovations in the present work are analytical and even conceptual, and that it is in order to appraise the value of the book by its specific results, such as we have extracted above; and of such let the author produce many more, by all means." The Swedish mathematician Arne Beurling, a specialist in harmonic analysis, muttered à propos distributions and Laurent Schwartz that "he has no uniqueness theorem." When Hörmander defended his thesis in 1955 (see below), Schwartz's student Jacques-Louis Lions was Hörmander's opponent, and his teacher Marcel Riesz, then 69 years of age, was present. Riesz, who had not read the thesis, feared that it would contain too much vague distribution terminology, but he was relieved by Lions's opening statement: " à la base il y a les inégalités de Hörmander."

The kernel theorem. Two important problems

Linear maps $f \to Kf$ are sometimes given by kernels $K(x,y)$ in such a way that

$$Kf(x) = \int K(x,y)f(y)dy, \quad x \in R^n, \, y \in R^m.$$

In particular, $K(x,y)$ may be a distribution on R^{n+m} and then K is a natural continuous map from C_0^∞ to $\mathscr{D}'$. One of the substantial results of distribution theory, Schwartz's kernel theorem (1950), says that the converse is true. The topological arguments of the original proof have later been greatly simplified.

Already at an early stage Laurent Schwartz could formulate two important problems in his theory: the existence of a fundamental solution for an arbitrary differential operator $P(D)$ with constant coefficients and division of a distribution by a polynomial $a(x)$, i.e. a solution of the equation $aS = T$ where T and S are distributions. Later Schwartz formulated the problem of characterizing hypoelliptic operators, i.e. those partial differential operators $P(D)$ with constant coefficients such that all distribution solutions u of $P(D)u = 0$ are infinitely differentiable. Both problems were solved within a decade.

Leon Ehrenpreis (1954) and Bernard Malgrange (1955) settled the first problem, Malgrange by a characteristic use of functional analysis which I will now sketch.

---

[2]See the chapter on Intrinsic hyperbolicity.

The first step is an inequality

$$|f(0)| \leq \frac{1}{2\pi} \int_0^{2\pi} |P(e^{i\theta})f(e^{i\theta})| d\theta$$

where $P(t) = t^m + \cdots$ is a polynomial of degree m and $f(t)$ is analytic in the unit disk. To see this, write $P(t) = \prod(t - a_k)$. Then the right side does not change if every factor $t - a_k$ with $|a_k| < 1$ is replaced by $1/(1 - \bar{a}_k t)$. The resulting integral without absolute values then equals $\prod' a_k f(0)$ where all a_k in the product have absolute values ≥ 1. Hence the inequality follows.

If now $P(t) = P(t_1, \ldots, t_m) = t_1^m + \cdots$ is a polynomial in several variables of degree m and $F(t)$ is the Fourier transform of the test function $\varphi(x)$, the inequality above means that

$$|F(t)| \leq \frac{1}{2\pi} \int_0^{2\pi} |(PF)(t_1 + e^{i\theta}, t_2, \ldots, t_n)| d\theta.$$

Then it is easy to verify that

$$|\varphi(0)| \leq \frac{1}{(2\pi)^n} \int_{R^n} |F(t)| dt \leq \varrho(P\varphi), \quad P = P(D),$$

for some permitted seminorm ϱ on $P(D)C_0^\infty$. Hence the Hahn-Banach theorem shows that the linear map $P\varphi \to P\varphi(0)$ extends to a distribution T such that $\varphi(0) = T(P\varphi)$, i.e. $P'T = \delta(x)$ so that T is a fundamental solution of the adjoint P'.

The fundamental solution constructed in this way is not tempered, and it is certainly hopeless to realize it as a function. The construction of a tempered fundamental solution is a special case of the problem of division: when $P(t) \neq 0$ is a polynomial, prove that $P(t)\varphi(t) \to \varphi(t)$ is a continuous map of tempered test functions. If so, $T(\varphi) = S(P\varphi)$ with a given tempered distribution T defines a temperered distribution S which extends to another such distribution U such that that $PU = T$. The continuity in question was proved by Hörmander (1958), and Lojasiewicz (1959) proved that division by a non-vanishing analytic function is locally possible for distributions.

Generalized functions

In Russia distributions were taught under the name of generalized functions in an attractive series of books by Gelfand and coauthors. In the first volume (1958, I) there is no hesitation in denoting distributions as functions by symbols like $f(x)$ and the object is to give an extensive series of examples. There are for instance explicit distributions representing $Q_+(x)^\lambda$ where λ is real and $Q(x)$ is any real quadratic form. A footnote (p. 118) has an audacious decomposition of the delta function

$$\delta(x) = (2\pi)^{-n} \int e^{i(x,\xi)} dx$$

into plane waves. Here $x = (x_1, \ldots, x_n)$ and $\xi = (\xi_1, \ldots, \xi_n)$ are dual real variables and $(x, \xi) = \sum x_k \xi_k$. Introducing polar coordinates $\xi = \varrho(\xi)\eta$ where $\varrho(\xi) > 0$ is homogeneous of degree 1, this gives

$$\delta(x) = \frac{(n-1)!}{(2\pi i)^n} \int_{\varrho=1} ((x, \eta) - i0)^{-n} \omega(\eta)$$

where $\omega(\xi)$ is an $n-1$-form defined by $\omega(\eta) \wedge d\varrho(\eta) = d\eta_1 \wedge \cdots \wedge d\eta_n$ when $\varrho(\eta) = 1$.

The second book (1958, II) in the series presents linear topology with the dictum that *Different problems require different linear spaces.* This leads to a number of linear spaces, for instance those called $K(M_p)$ whose elements are infinitely differentiable functions φ with seminorms

$$\sup M_p(x) |D^\alpha \varphi(x)|, \quad |\alpha| \le p,$$

where the weights $1 \le M_0(x) \le M_1(x)$ are continuous and may be $+\infty$ for large x. The same seminorms applied to analytic functions give a space $Z(M_p)$. Schwartz's space $\mathscr{S}$ may be modified in various ways and there are the spaces W where all $\varphi^{(p)}$ or $z^p \varphi(z)$ have exponential bounds. In many cases, the Fourier transform maps one space of test functions into another so that the generalized Fourier transform maps one dual space of generalized functions into another. In (1958, III) some of these ideas are applied to produce uniqueness results, similar to the classical ones for the heat equations, for initial value problems for partial differential equations and systems with constant coefficients.

Volume IV (1960) treats generalized functions in Hilbert space. It starts with spectral theory and ends with a theory of Gaussian measures and their Fourier transforms. The connecting link is the theory of rigged Hilbert spaces which is a triple $\Phi \subset H \subset \Phi'$ where H is a Hilbert space containing a continuously and densely imbedded linear space Φ and contained in its dual Φ' in such a way that the inner product of H serves as the duality between Φ and Φ'. This construction is very useful in various situations when H has generalized eigenfunctions, but the term 'rigged Hilbert space' had only a short life. The fifth volume (1962, V), finally, has a less clear connection with generalized functions. It deals with integral geometry, for instance the Radon transform, and unitary representations of the Lorentz group.

This very interesting series of books has not had much influence in the West, but it signals a development where analysis has absorbed so much of the theory of linear spaces and their topology that it needs. This is reflected in the thinking of the generation of mathematicians who grew up with distributions. For them functional analysis and distributions are auxiliary things to be taken for granted.

Distributions and general partial differential operators

Looking back, the most notable results in the 1940's about high order differential equations were Herglotz's constructions of fundamental solutions of strongly hyperbolic differential operators, Petrovskii's proof of the analyticity of solutions of elliptic equations with analytic coefficients and his existence proof for Cauchy's problem for strongly hyperbolic systems. At the time Schwartz took a lively interest in these results because he thought that distributions would play a big part in a future theory of general partial differential operators. He has been justified by the later development.

The theses by Malgrange (1955) and Hörmander (1955) were the first comprehensive treatments of this field. Malgrange used distributions and Schwartz's notation systematically and devoted himself also to convolution equations. Hörmander used distributions quite sparsely. Instead, square integrable functions were extensively employed. The main part deals with domains and ranges of partial differential operators $P(D)$ with derivatives $D = \partial/i\partial x$ in several variables $x = (x_1, \ldots, x_n)$ and characteristic polynomials $P(\xi)$. There is a minimal domain in a bounded open set Ω consisting of the closure in $L^2(\Omega) \times L^2(\Omega)$ of the pairs $f, P(D)f$, $f \in C_0^\infty$, and a maximal domain consisting of all $f \in L^2(\Omega)$ such that $P(D)f$, taken as a distribution, belongs to $L^2(\Omega)$. Using an inequality by Malgrange (see above) it is shown that every $P(D) \neq 0$ is invertible on its minimal domain. On the other hand, the maximal domain seems to characterize the operator. Operators whose maximal domains permit multiplication by smooth functions are said to be local. An equivalent condition for this was found expressed in terms of the characteristic polynomial or symbol $P(\xi)$. If this polynomial is complete, i.e. depends on all variables, this condition is necessary and sufficient for $P(D)$ to be hypoelliptic, i.e. all solutions u of $P(D)u = 0$ are infinitely differentiable.

The last part deals with operators $P(D)$ of principal type, those for which the gradient of the principal part of $P(\xi)$ never vanishes. By a tour de force it is proved that if $P(x, D)$ has smooth coefficients, if $P(x, \xi)$ has degree m and is real of principal type near the origin, then

$$\int_\Omega \sum_{|\alpha| < m} |D^\alpha u(x)|^2 dx \leq \text{const} \int |P(x, D)u(x)| dx$$

when $f \in C_0^\infty(\Omega)$ and Ω is sufficiently close to the origin. From this follows in the usual way the local solvability of the equation $P^*(x, D)u = v \in L^2$.

In Hörmander's comprehensive book (1963) time was ripe for the systematic use of distributions in the theory of partial differential operators. It starts with a terse chapter on distribution theory and in the sequel the notions introduced are used as a matter of course. Here distributions are used as a natural tool and the problems are the classical ones of the structure of solutions and existence and uniqueness of various boundary problems.

One central section continues one of the problems of the thesis, the surjectivity of operators of principal type. It turned out that the previous result (see above) is not true for such operators unless the principal types are real or else satisfy special conditions. Work in this area gave Hörmander a Fields medal in 1962 at the International Congress in Stockholm.

In physics distributions started to be taught in an elementary way. They had a great importance in the quantum field theory where the field operators are highly singular and achieve mathematical sense as distributions (see Wightman and Gårding (1965) and Wightman's historical review (1996)).

Hyperfunctions

As generalized functions, the distributions are a proper subset of the hyperfunctions whose theory originated in two papers (1959) by Mikio Sato. The aim is that of distributions but otherwise there seems to be no connection. A hyperfunction on an open interval I of the real axis is an equivalence class of functions f^+ and f^-, analytic in two open sets O^+ and O^- in the upper and lower half-planes adjoining I, modulo functions which are analytic across I. Operations by differential operators with analytic coefficients on the pair f^+, f^- are then transferred to the hyperfunctions. When $f^+(z)$ and $f^-(z)$ do not grow more than an inverse power of $|\operatorname{Im} z|$ as z approaches compact parts of I, the hyperfunction can be identified with a distribution in the interval I, otherwise not.

The definition above is marred by the dependence on the choice of the adjoining open sets, but this difficulty can be eliminated and the existence of a support for hyperfunctions established by the use of sheaf theory, a tool which has come to pervade the very rich later developments in several variables (see Martineau (1960), Sato, Kawai, Kashiwara (1973)).

Hörmander (1983a) replaced the complex variables by a harmonic transform in one extra variable and found an alternative definition closer to that of distributions as follows. Consider first the space $A'(K)$ of analytic functionals T on a compact subset K of R^n, defined as linear maps from the space of entire analytic functions $\varphi(z)$ such that

$$|T(\varphi)| \leq C_\omega \sup_\omega |\varphi|$$

for every open neighborhood ω of K. If $X \subset R^n$ is open and bounded, the space $B(X)$ of hyperfunctions on X is then the quotient $A'(X)/A'(\partial X)$.

We shall not go further in the theory, but hyperfunctions are important in microlocal analysis, which is the subject of the last section.

Pseudodifferential operators

The theory of pseudodifferential operators originated as the use of singular integrals by Calderón and Zygmund (1957) with roots in the works by Mihlin and others. In a later development, spurred by the interest in the index of elliptic systems, the theory took another form as realizations of operators of the form $P(x, D)$, $D = \partial/i\partial$, for which, unlike partial differential operators, the characteristic polynomial $P(x, \xi)$ is no longer a polynomial (see Kohn and Nirenberg (1965) and Hörmander (1965)). It should then have properties which make sense of the formula

$$(1) \qquad P(x, D)f(x) = (2\pi)^{-n} \int e^{i(x,\xi)} P(x, \xi) F(\xi) d\xi$$

where

$$(2) \qquad F(\xi) = \int e^{-i(x,\xi)} f(x) dx$$

is the Fourier transform of a function or distribution f and the left side should be at least a distribution. To fit distribution theory, $P(x, \xi)$ should then be a C^∞ function of x and at the same time a multiplier of $\mathscr{S}$ and $\mathscr{S}'$ as a function of ξ, i.e. it should have bounded derivatives of all orders. A simple way of achieving this goal is to require that

$$P(x, \xi) \sim \sum_m^{-\infty} P_k(x, \xi)$$

is an asymptotic sum of terms P_k which are homogeneous of degree k for large values of ξ. The cost of this construction is that $P(x, D)$ is an operator modulo C^∞ functions. The benefit is a symbolic calculus: if $P_m(x, \xi) \neq 0$ is the principal part of $P(x, D)$, the principal part of a product $R(x, D) = P(x, D)Q(x, D)$ is the product of the principal parts. More precisely, the asymptotic series of $R(x, D)$ can be computed from those of $P(x, D)$ and $Q(x, D)$.

If we undress the Fourier transform in (1), a pseudodifferential operator appears as a map with the distribution kernel

$$K(x,y) = (2\pi)^{-n} \int e^{i(x-y,\xi)} P(x,\xi) d\xi.$$

Integrations by parts when $x \neq y$ show that $K(x,y)$ is an indefinitely differentiable function when $x \neq y$.

It should be clear from the above that the theory of pseudodifferential operators of this form is a secondary impact of distribution theory in analysis. The seminal papers on pseudodifferential operators in the mid sixties have been followed by many others where pseudodifferential operators have proved their usefulness and served as a frame for some very important analysis.

It is not difficult to extend the essential features of distributions to C^∞ manifolds, and with this the theory of pseudodifferential operators also extends. This has made it possible to create a suitable analytic framework for the index theory of elliptic systems on a manifold.

Microlocal analysis

It is an elementary fact that the singularities of distributions turn up in the high frequency behavior of their Fourier transforms. Microlocal analysis is a study of the high frequency structure of singularities of functions and distributions. A beginning was made by Maslov (1965). Microlocal analysis exists also for hyperfunctions and was invented by Sato (1959), 11 years before Hörmander (1970) could announce the beginnings of microanalysis for distributions where it appears as a second order impact of this theory. It uses the notion of distribution and the notion of Fourier transform of a distribution with compact support, both introduced and advocated by Laurent Schwartz. Only a sketch can be given here.[3]

In the final analysis, the setting of microlocal analysis is the cotangent bundle $T^*(X)$ of a differentiable manifold X with local coordinates x, ξ and invariant differential form $\omega = (dx, \xi)$. But the theory has to start in R^n.

To make the Fourier transform reasonably invariant under linear changes of variables, a linear change of variables in the x variable should carry with it the inverse adjoint change of the ξ variables, making the interior product (x, ξ) invariant. With this understanding (x, ξ) is an invariant form on the cotangent bundle T^* of R^n.

Let u be a distribution on R^n and let $f \in C_0^\infty$. Simple arguments show that the growth at infinity of the Fourier transform $v(\xi)$ of fu gets smaller in all directions when f is replaced by a product fg and $g \in C_0^\infty$. Hence, as the support of f shrinks to a point x, the union of the open conical sets in ξ-space on which $v(\xi) = O(|\xi|^{-N})$ for all large ξ and large N has a complement $C(x)$, the origin not included. The members of $C(x)$ are the high frequencies which are necessary to reconstruct the singularity of u close to x. When $C(x)$ is empty, the distribution u is an infinitely differentiable function close to x. The union of the sets $C(x)$ is a closed subset of T^* called the wave front set $\mathrm{WF}(u)$ of u in which $C(x)$ is the fibre over x. The

---

[3]For a full exposition, see Hörmander's monumental four volumes (1983–1985) on the analysis of linear partial differential operators. For hyperfunctions there is a corresponding development (see Schapira (1977, 1981)) which will not be the subject here.

projection of $WF(u)$ on x-space is the singular support of u, i.e. the complement of the open set where u is an infinitely differentiable function.

The wave front sets of the one-dimensional distributions $\delta(x)$ and $1/(x \pm i0)$ are the simplest examples. The fibres are empty over $x \neq 0$ and the fibres over 0 are $(-1) \cup (1)$, (1) and (-1) where ± 1 stands for the positive or negative real axis.

When $P(x, D)$ is a pseudodifferential operator and u is a distribution, then $WF(P(x,D)u)$ is a subset of $WF(u)$. This follows from the fact that the kernel of the operator is smooth outside of the diagonal.

If a distribution $u(x)$ is independent of x_1, its Fourier transform is supported in $\xi_1 = 0$ and independent of the value of x_1. This is the simplest example of Hörmander's basic propagation of singularities theorem (1970) which runs as follows: If $P(x, D)$ is a pseudodifferential operator of real principal type in the sense that the gradient of its principal symbol $P_m(x, \xi)$ is real and does not vanish for $\xi \neq 0$, and $P(x, D)u \in C^\infty$, then $WF(u)$ is invariant under canonical flows defined by

$$dx/dt = \partial P_m(x, \xi)/\partial \xi, \quad d\xi/dt = -\partial P_m(x, \xi)/\partial x.$$

The principal symbol $P_m(x, \xi)$ is invariant under this map and so is the canonical differential form (ξ, dx) when $P_m(x, \xi) = 0$.

With this result, microlocal analysis made a surprising contact with Hamiltonian mechanics and the classical theory of the characteristics of first order partial differential equations. Therefore the orbits above are aptly named (null) bicharacteristics.

The notion of wave front set is very relevant when applied to oscillatory integrals of the following general form:

$$(3) \qquad\qquad u(x) = \int a(x, \theta) e^{i\lambda(x, \theta)} d\theta,$$

extensively studied by Hörmander (1971). The amplitude $a(x, \theta)$ is supposed to be a smooth function with x in some open subset of R^n and $\theta \in R^N$. It is assumed that $\partial_\theta^\alpha \partial_x^\beta a(x, \theta) = O(|\theta|^{m-|\alpha|})$ for large θ, locally uniformly in x. The phase function $\lambda(x, \theta)$ is supposed to be smooth and real and homogeneous of degree 1 in θ. The assumption that $d\lambda \neq 0$ makes u a distribution which is a smooth function of x unless $\lambda_\theta(x, \theta) = 0$ for some θ. In practice, the amplitude $a(x, \theta)$ is often polyhomogeneous, i.e. an asymptotic sum of terms of decreasing integral homogeneity in θ as described above for pseudodifferential operators.

The wave front set of the oscillatory integral above is contained in the set of pairs x, λ_x such that $\lambda_\theta(x, \theta) = 0$. When the phase function is regular, i.e. the differentials $d\lambda_\theta$ are linearly independent, this equation defines a conical Lagrangian manifold, a submanifold of $T^*(R^n)$ of maximal dimension on which the differential form (ξ, dx) vanishes. One important result of Hörmander (1971) is that two oscillatory integrals with regular phase functions with the same Lagrangian conical manifold produce the same distributions modulo smooth functions, at least when the conical supports of the amplitudes are small.

When the phase function λ of (3) has the form $\lambda(x, y, \theta), x \in R^n, y \in R^m$, the integral $I(x, y)$ represents the kernel of what is called a Fourier integral operator (Hörmander (1971)). Generally speaking, by a theorem by Egorov (1969), the corresponding operator will map distributions u to distributions v such that

$$WF(v) \subset C(WF(u)).$$

Here $C = (x, \xi, y, -\eta)$ is a canonical relation such that (x, ξ, y, η) belongs to the Lagrangian manifold defined by I. This fact makes Fourier integral operators a powerful tool of microlocal analysis which permits a change of variables in the cotangent bundle mixing its two ingredients.

The important results sketched above have been the starting points of an extensive theory of propagation and reflection of singularities of solutions of linear and also non-linear partial differential equations. This development is a long range effect of the ideas of distribution theory. No details can be given here and the reader is referred to Hörmander's series (1983a, b), (1985a, b).

Bibliography

BANACH S.

1932. *Théorie des opérations linéaires*, Monografje Matematyczne, Warszawa, 1932.

BOCHNER S.

1932. *Vorlesungen über Fouriersche Integrale*, Leipzig, 1932.

1952. *Théorie de distributions*, Bull. Am. Soc. **58** (1952), 78–85.

CALDERON A. P., ZYGMUND A.

1957. *Singular integrals and differential equations*, Amer. J. Math. **79** (1957), 901–921.

CARLEMAN T.

1944. *L'intégrale de Fourier et les questions qui s'y rattachent*, Publ. Sci. Inst. Mittag-Leffler, Uppsala, 1944.

EGOROV Ju. V.

1969. *The canonical transformations of pseudodifferential operators*, Usp. Mat. Nauk **24** (1969), no. 5, 235–236.

EHRENPREIS L.

1954. *Solutions of some problems of division*, I, Amer. J. Math. **76** (1954), 883–903.

FRIEDRICHS K.

1944. *The identity of weak and strong extensions of differential operators*, Trans. Am. Math. Soc. **55** (1944), 132–151.

GELFAND I., SHILOV G. E.

1958. *Generalized functions, I (Generalized functions and operations on them), II (Spaces of test functions and generalized functions), III (Some problems in the theory of differential equations)*, Moscow, 1958.

GELFAND I., VILENKIN N. Ya.

1960. *Some examples of harmonic analysis. Rigged Hilbert spaces*, Moscow, 1960.

GELFAND I., GRAEV M. I., VILENKIN N. Ya.

1962. *Integral geometry and related questions. The theory of representations*, Moscow, 1962.

HÖRMANDER L.

1955. *On the theory of general partial differential operators*, Acta Math. **94** (1955), 161–248.

1958. *On the division of distributions by polynomials*, Arkiv f. Mat. **3** (1958), 555–568.

1963. *Linear partial differential operators*, Springer, 1963.

1965. *Pseudodifferential operators*, Comm. Pure Appl. Math. **18** (1965), 501–507.

1970. *Linear differential operators*, Actes Int. Congr. Math. Nice, 1970.

1971. *Fourier integral operators I*, Acta Math. **127** (1971), 79–183.

1983. a. *The analysis of linear partial differential operators*, I, Springer, 1983.

1983. b. *The analysis of linear partial differential operators*, II, Springer, 1983.

1985. a. *The analysis of linear partial differential operators*, III, Springer, 1985.

1985. b. *The analysis of linear partial differential operators*, IV, Springer, 1985.

KOHN J. J., NIRENBERG L.

1965. *On the algebra of pseudodifferential operators*, Comm. Pure Appl. Math. **18** (1965), 269–305.

ŁOJASIEWICZ Ya. B.

1959. *Sur le problème de division*, Studia Math. **18** (1959), 87–136.

MALGRANGE B.
  1955. *Existence et approximation des solutions des équations aux dérivées partielles et des équations de convolution*, Ann. Inst. Fourier (Grenoble) **6** (1955–56), 283–306.
MARTINEAU A.
  1960. *Les hyperfonctions de M. Sato*, Sém. Bourbaki 1960–61, Exposé no. 214.
MASLOV V. P.
  1965. *Theory of perturbations and symptotic methods*, Izd. Mosk. Gos. Univ., Moscow, 1965.
MIHLIN S. G.
  1956. *On the multipliers of Fourier integrals*, Dokl. Ak. Nauk SSSR **109** (1956), 701–703.
SATO M.
  1959. *Theory of hyperfunctions*, J. Fac. Sci. Univ. Tokyo, Sect. I, **8** (1959–60), 139–193, 387–436.
SATO M., KAWAI T., KASHIWARA M.
  1973. *Hyperfunctions and pseudo-differential functions*, Lecture Notes in Mathematics, vol. 287, Proceedings of a conference at Katata 1971.
SCHAPIRA P.
  1977. *Propagation at the boundary and reflection of analytic singularities of solutions of linear partial differential equations*, Publ. RIMS Kyoto Univ. **12** Suppl. (1977), 441–453.
  1981. *Propagation at the boundary of analytic singularities*, Nato Adv. Study Inst. on Sing. in Bound. Value Problems, Reidel Publ Co., Doordrecht, 1981, pp. 185–212.
SCHWARTZ L.
  1945. *Généralisation de la notion de fonction, de transformation de Fourier et applications mathématiques et physiques*, Ann. Univ. Grenoble **21** (1945), 57–74.
  1950. *Théorie des noyaux*, Proc. Int. Congr. Math. Cambridge, 1950, pp. 220–230.
  1951. *Théorie de distributions*, I, II, Paris, 1950–51.
  1952. *Transformation de Laplace des distributions*, Comm. Sém. Math. Univ. Lund, Tome suppl. dédié à Marcel Riesz (1952), 196–206.
SOBOLEV S.
  1936. *Méthode nouvelle à résoudre le problème de Cauchy pour les équations linéaires hyperboliques normales*, Mat. Sb. **43** (N.S. **1**) (1936), 39–72.
TITCHMARSH T. C.
  1937. *Introduction to the theory of Fourier integrals*, Oxford, 1937.
WEYL H.
  1940. *The method of orthogonal projection in potential theory*, Duke Math. J. **7** (1940), 411–444.
WIENER N.
  1926. *The operational calculus*, Math. Ann. **95** (1926), 158–189.
  1932. *Tauberian theorems*, Ann. of Math. **33** (1932), 1–100.
WIGHTMAN A., GÅRDING L.
  1965. *Fields as operator-valued distributions in relativistic quantum theory*, Arkiv f. fysik **28** (1965), 149–189.
WIGHTMAN A.
  1996. *How it was learned that quantized fields are operator-valued distributions*, Fortschr. Phys. **2** (1996), 143–178.
ZYGMUND A.
  1935. *Trigonometrical series*, Dover Publ., 1955.